PRAISE FOR *REWIRED*

"Moving an arm involves probably 500 million neurons. The idea to translate thoughts into motion had not been done before. Dr. Ajay Seth's groundbreaking surgery has caused prosthetic research to leap forward. Dr. Seth has caused his patient Melissa Loomis to be the most advanced amputee in the world."

—Mike McLaughlin
Chief Biomedical Engineer/Prosthetics; Chief Engineer for Research and Exploratory Development, Johns Hopkins Applied Physics Laboratory

"Dr. Ajay Seth is a critical part of our SWAT team as the volunteer on-site medical advisor. We never know when we'll need to call on him, but when we need Dr. Seth, he always responds. Our guys get fantastic medical backup because he is civic- and community-minded, donating his time to the police department."

—Donny Miller
SWAT Commander, Canton, Ohio

"Dr. Ajay Seth is always on his game on sports radio. He knows a lot about basketball, and it shows, but his knowledge of medicine is so vast, it's hard to imagine. You can see it in his ability to care for others, including professional athletes."

—Keith McLeod
Former NBA player; varsity basketball coach, GlenOak High School, Canton, Ohio

"I work with Dr. Ajay Seth in medical equipment and have a firsthand look at the way he cares for patients. As a former professional athlete, I've been in a lot of doctors' offices, but Dr. Seth's level of expertise proves he's one of the best."

—Joe Inglett
Former MLB player and medical representative

"Being raised in a sports family, I have been surrounded by professional athletes, and their doctors, my entire life. Dr. Seth is the kind of doctor who gives special care to every individual. The community depends on his volunteer work with kids in school athletic programs. My wife and I count on him for our family's health, and he always comes through."

—Todd Blackledge

Former NFL quarterback; basketball coach, Hoover High School, North Canton, Ohio

"Dr. Ajay Seth is the best sports doctor of any I've ever worked with in three decades of sports broadcasting. I appreciate him volunteering his talents as a media expert for orthopedic problems on our show."

—Kenny Roda

The Kenny Roda Show, "Cleveland's Best Sports Talker"

rewired

rewired

AN UNLIKELY DOCTOR, A BRAVE AMPUTEE, AND THE MEDICAL MIRACLE THAT MADE HISTORY

AJAY K. SETH, MD

W Publishing Group

An Imprint of Thomas Nelson

© 2019 Dr. Ajay Seth

All rights reserved. No portion of this book may be reproduced, stored in a retrieval system, or transmitted in any form or by any means—electronic, mechanical, photocopy, recording, scanning, or other—except for brief quotations in critical reviews or articles, without the prior written permission of the publisher.

Published in Nashville, Tennessee, by W Publishing Group, an imprint of Thomas Nelson.

Thomas Nelson titles may be purchased in bulk for educational, business, fund-raising, or sales promotional use. For information, please email SpecialMarkets@ThomasNelson.com.

The G. K. Chesterton quote in chapter 8 is taken from "The Eternal Revolution," in *The Collected Works of G. K. Chesterton, Volume 1* (San Francisco: Ignatius Press, 1986), 325.

The children's book referenced and quoted in chapter 39 is Kobi Yamada and Mae Besom, illustrator, *What Do You Do with an Idea?* (Seattle: Compendium Inc., 2014).

Any Internet addresses, phone numbers, or company or product information printed in this book are offered as a resource and are not intended in any way to be or to imply an endorsement by Thomas Nelson, nor does Thomas Nelson vouch for the existence, content, or services of these sites, phone numbers, companies, or products beyond the life of this book.

ISBN 978-0-7852-2124-1 (TP)
ISBN 978-0-7852-2113-5 (HC)
ISBN 978-0-7852-2119-7 (eBook)
ISBN 978-0-7852-2125-8 (Audio)

Library of Congress Cataloging-in-Publication Data

Library of Congress Control Number: 2018030232

Printed in the United States of America

22 23 24 25 26 LSC 10 9 8 7 6 5 4 3 2 1

For my parents, Raj and Asha Seth, who immigrated from India for the sole purpose of providing their children a better education and the chance for unlimited opportunities. I hope I have fulfilled their dreams.

For my wife, Kimberly; my children, Jaideep and Trinity; and my sister, Angela. I hope I've been able to set an example that it doesn't matter what you do in life; it's about the pursuit of accomplishing the impossible.

For Melissa, who believed and trusted in me without any hesitation or doubt.

CONTENTS

PROLOGUE

The idea for this book began, of all places, under a basketball hoop in North Canton, Ohio. A friend and I were chatting as we watched our kids play.

"So tell me about that operation," he said. "The one I saw on the news. I'm trying to wrap my mind around it."

"I get that a lot." I smiled and did my best, one more time, to distill a complicated story. As always, I stressed the truly amazing stuff, which I was still trying to wrap *my* mind around.

My friend listened carefully and said, "You know what? All of that gets me thinking. I mean, I'm not the super-religious type. But when I hear that story, it's hard not to believe a higher power is at work. You ought to write a book so everybody can hear about this. I'll buy the first copy—and maybe the second."

Everywhere I went—to the grocery store, out to eat—I was expected to tell this story one more time. It seemed to captivate everyone who heard it. So in the end I decided a book made a lot of sense.

This book is about Melissa Loomis, a forty-three-year-old woman who suffered from one of the worst infections I've ever seen. She endured four months of surgeries—fourteen total—as well as gut-wrenching decisions and emotional ups and downs. In that way she's like many other patients.

But *just* in that way. Otherwise, she's utterly one of a kind.

I've dealt with patients in many situations. What makes Melissa Loomis different is her unique combination of thoughtfulness, kindness, eerie calm, and (especially) courage: she has never given in to self-pity or held a sense of entitlement, even to something as basic and human as having two functional arms.

She didn't stumble into this narrative by chance. I believe she was *chosen* for it, and after you've read this narrative, you just might too. Melissa was the perfect choice for the trial no one saw coming.

But the events themselves were equally singular. As you'll see, we saw twist upon twist, until the word *coincidence* simply no longer fit into its grand design. Still, I didn't see the whole picture for quite a while. One day early in February 2016, my mind assembled all the pieces and saw the full panorama of greater things at work.

If I can show you that picture clearly, you'll see two things: (1) a portrait of the kind of person who prevails in the face of life's worst moments—someone like Melissa, with her positive, rugged, and unflappable spirit—and on top of that (2) a clear picture of everyday life touched by the divine. I attribute 51 percent of the result of the surgery to God and 49 percent to Melissa herself.

Neither force is something I can explain.

What about me? I'm just here to tell the story. Yes, I performed the surgery, but most surgeons could have done it too. I tell these things in the hope that you'll see both what quiet heroism looks like and how we know God still moves mountains every now and then. If this story transforms your understanding of either, then this narrative will have been valuable. If it changes the way you see both, then that's just another miracle—as it has been for me and for so many others.

◆◆◆

Before I begin, I hope you'll indulge my need to express gratitude in corners where it's past due. My appreciation begins with my parents, who immigrated to the United States from India for the sole purpose of finding a better life for their children. They sought a land of opportunity, and in so doing they blessed their kids beyond measure.

My sister has loved me and supported me at every step. From her I learned the lesson of pushing back when life closes in, never giving in or giving up.

Then there are my wonderful wife and kids: always loving, always supporting, always putting up with me. With family like mine and a patient like Melissa, I pretty well started this inning of the game at third base. Still, we all needed the grace and power of God moving in ways beyond our comprehension.

That, of course, is the higher power my friend under the basketball hoop mentioned. How can I express my gratitude to God? This is all for his glory, never mine. There's nothing remotely special about me. Just as I reach for certain tools in surgery, he reached for me, his implement in some divine operation. I'm happy to be his scalpel, his forceps, or whatever use he might have for me.

Speaking of which, are you ready for this operation? Generally I sedate my patients at this point. But this is one time I hope not to put the patient to sleep! Turn the page, and I'll introduce you to a curious and fateful raccoon, as well as to Melissa, an absolutely ordinary woman who emerged from a suburb of North Canton, Ohio, with an aching arm and a world-changing destiny.

—Ajay Seth, MD

PART 1
THE RACCOON

1

BITE-SIZE

Just an ordinary raccoon—that's what you'd think at first. I know I did.

I couldn't have been more wrong; this was a special raccoon. This one set in motion the strange wheels of destiny, ultimately changing the world. The furry, black-masked character was in and out of our story in no more than five minutes. But he left a mark—two marks, actually, on my patient's right wrist. And one colossal mark on medical history.

This is a story filled with twists and turns, ups and downs, many people, diverse locations, happy and sad moments. But never forget that little raccoon who started it all.

It was six o'clock on the morning of July 25, 2015, when the raccoon furtively emerged from the shadows. He was wandering through the woods of Stark County, Ohio, looking for breakfast. Before him sat a grassy clearing, a promising one, and a four-foot wire fence enclosing it. He scampered to the top of the fence, prepared to drop to the other side—and froze in place.

Two raucous, yapping dogs had sounded the alarm.

The canine couple went by the names of Vivian Sue Dog and Vincent Van Dog. They were owned and adored by their human caretaker, Melissa Loomis. They loved her, too, but at the moment they were in a barking frenzy over discovering an intruder into their little world.

The two pets had gently awakened their owner a few minutes ago, as usual, by pushing their heads firmly against her bed. It was time for their morning romp in the wide world of the backyard. Melissa had tucked her toes into her slippers, yawned, and followed Vivian and Vincent to the door. The wire fence limited their travels, but there was still enough room for the dogs to stretch their legs, attend to crucial body functions, and sniff around for anything new and interesting in the way of critters—such as, in this case, a raccoon.

Squirrels and chipmunks were pretty much par for the course. But a raccoon? Well, that was a special visitor. This was a special day, and the dogs were lifting their voices to let the world know.

Melissa quickly hurried to them. She normally waited on the porch, scrolling through Facebook posts on her phone. But five minutes into their romp, she'd heard a hissing followed by furious barking. The neighbors were going to love that. Putting away her phone, she came running and saw the nature of the standoff.

A raccoon clung to the fence. Not a cute, fuzzy cartoon character as drawn by Disney. This one looked feral and desperate as the dogs yapped furiously, growled, and closed in on their prey. Vivian showed especially deadly intent; she was on the verge of launching her attack.

You might say Vivian Sue isn't the name for a savage, bloodthirsty predator, but Melissa would reply that you just don't know Vivian Sue Dog. She's loving but deadly serious when it comes to unwanted varmints. I once asked Melissa what breeds her dogs were. She told me Vincent is a beagle, and she believes Vivian is a Treeing Walker Coonhound.

Well, no wonder!

The raccoon was sitting on the fence for the moment, literally and figuratively, but Melissa was certain it wasn't going down without a fight; he looked ready to take the battle to the dogs, tooth and nail.

Melissa was particularly worried about the raccoon, who wasn't going to stand much chance against two large dogs with hunting in their blood. But what were her options? She could go back into the house and get Neil, her husband. She could find a rock and throw it, hopefully chasing the animal back into the woods.

Neither of those would be timely enough. Things were about to go sideways. *To heck with it*, Melissa thought and came at the raccoon herself. She began waving her arms and shouting, "Shoo! Shoo! Get out of here, raccoon! Go away!"

I had to stop Melissa right there when she first told me the story.

"Really?" I asked. "Were you hoping the raccoon spoke English?"

"No, but—"

"And if savage, teeth-baring dogs couldn't scare the animal away, did you figure your birdlike, flapping arms would do the trick?"

Melissa just smiled and shrugged. It wouldn't have occurred to her to back down. She was worried about that raccoon, you see. Melissa has a heart for animals.

Things indeed went sideways. When the raccoon curiously eyed the strange woman, the distraction was just the opening Vivian Sue Dog had been waiting for. She sprang at the fence, clamped her teeth on the raccoon's tail, and flung the animal onto the ground, where the dogs could settle the issue.

Melissa had to do something and quick. My suggestion would have been, "Get the heck out of the line of fire!" Instead, she grabbed a dog collar with each hand, in hopes of dragging her pets back toward the house.

And that's the very moment when life for Melissa Loomis came to a fork in the road; she chose a hard left and accelerated down that unknown path with no chance of turning back—because the frantic raccoon jumped onto Melissa's right arm, sank his teeth into the bottom of her right wrist, and scratched her left arm in the bargain. Melissa grabbed the creature with both hands and heaved him over the fence.

Exit the raccoon, his role in this story complete. He vanished into the line of trees and the mists of uncertainty. Not so for Melissa. Her story had just begun.

She suddenly realized she was in pain—*boy*, was she in pain! She let out a serious howl. The dogs, their plans plummeted as they watched their hoped-for target flee into the woods, turned and regarded her with concern.

Melissa's yelp carried into the house and awakened Neil, who quickly ran to the back door, threw it open, and took in a sight he would never forget: his wife walking toward him with her arms thrust out before her, blood flowing from both.

I'll be honest. I hate the sight of blood—I hate it a *lot*—and my personal pain threshold isn't too impressive. And yes, this is a surgeon telling you that: somebody who performs, well, surgery, and deals with the sight of blood as a job description. It's true. When my kids skin a knee or lose a tooth, I don't handle it well. But if I'd been Melissa *or* Neil in this scenario, I'd probably have gone to pieces.

This was Melissa, however. She'd already put in her one good bellow and gotten it out of her system. Now she merely walked toward her husband and said, "I just lost an argument with a raccoon."

Sure, Melissa had been attacked by a feral animal, she had an arm throbbing with pain, and what was she up to? Tossing out a wry one-liner.

Ladies and gentlemen, Melissa Loomis.

Neil, a bit more practical at the moment, thought, *This isn't good*—at all. *We need to do something—right now!* All he knew was that his wife had been bitten and scratched by a raccoon. Was it rabid? Did it carry some kind of disease?

There was another factor to consider, and it added to Neil's concern. Melissa is a diabetic, making her risk of infection from a bite much higher than it would be for someone who didn't suffer from diabetes.

"We've got to get you to the emergency room, but we'd better do what we can first," Neil said. He guided her inside, where he helped clean her wounds, including the scratches, and bandaged her wrist with gauze.

Melissa felt it was a shame and a waste of time to be sitting in the waiting room of an ER on such a pretty summer morning. Her husband was eyeing those bites and taking the whole thing seriously. The dogs were at home, still trying to calm down. And somewhere there was a raccoon telling his forest friends that he'd just lived out the furry version of a horror movie.

But for Melissa, it wasn't a big deal. She took her phone to the medical facility so she could resume reading posts from her friends on Facebook. She had no idea that the wild ride of a lifetime was about to pick up speed.

2
PROTOCOL

Fifty-four minutes.

Melissa kept track. She and her husband arrived at the ER fifty-four minutes after the raccoon sunk his choppers into her wrist.

They parked the car, filled out the usual forms, and saw an ER physician. He quickly took note that Melissa had diabetes. Diabetics lack the strength other individuals might have to fight off an infection, so physicians are very careful.

The attending physician decided to start Melissa on a rabies protocol and a regimen of oral antibiotics. The gauze was removed, the wounds were cleaned a bit more thoroughly, Melissa's wrist was dressed with sterile bandages, and she was instructed when to return for three more shots to finish off her rabies vaccine, while taking her oral antibiotics each day. Rabies inoculation and common antibiotics would normally end the story, so the patient was discharged.

You or I would have been ready to go home and get a little rest after all this excitement and pain. Who needs the daily hassles of a job? But Melissa went to her office, with her right arm wrapped. She simply couldn't see any good reason to "lie around the house." She worked as a food services director, and she knew there were menus to be prepared and supplies to be ordered.

"My blood sugars were stable," she pointed out later. "I didn't miss a day of work." She didn't say it proudly, just as a fact. She even attended a fund-raiser.

A few days passed and then a few more. The oral antibiotics weren't making much headway, the pain wasn't going away, and her arm was still red. She walked into our office, consulted with one of our doctors, and was immediately sent to the hospital for a more rugged round of antibiotics—this time to be administered intravenously.

"The hospital? Really?" Melissa asked. It seemed like overkill to her. Her arm was hurting, and that was an inconvenience, but things like this healed up quickly, didn't they? Surely if she had to check in to a hospital, she'd be in and out after twenty-four hours. This would be the first time she'd miss a day of work.

On the way she called her sister, Michelle Dalpra. "Can you believe it?" she asked. "They're going to put me on an IV. For a silly animal bite."

Michelle didn't like the sound of it, but she had to agree with her sister's good sense. How serious could something like this be? There are amazing medicines and doctors. This wasn't a big deal.

While all this was happening, I was at a North Canton, Ohio, high school football field. I volunteer as a team doctor at Hoover High School for all its athletic teams, and the football team was into heavy practices for the coming season. Nobody was suffering from heat exhaustion or a tweaked ankle, so I was just hanging loose and having a good time.

My cell phone rang. I glanced at the phone, and the caller ID said "Tyler Crockett"—one of my nurses, the guy who always knew where to find me.

"What's up, Tyler?" I said.

"We've got a forty-three-year-old lady who was bitten by a raccoon, Dr. Seth. We need you to admit her."

"Raccoon? Really? That must be some kind of bad bite."

"She's been on antibiotics, and they've made no headway. She's diabetic—that may be part of the problem. She needs to be in for an IV round."

"Gotcha. We'll try that for twenty-four hours, and that should tell us where we are."

It was normal protocol for an animal bite infection. You hope that intravenous antibiotics, which go directly into the bloodstream, will get the infection under control, and then you won't have to consider surgery.

"Thanks, Dr. Seth."

"Later."

And that's how it began for me—nothing out of the routine, even for an animal bite. At this point in my medical career, I'd handled hundreds of infections. This kind of case pointed toward a forty-eight-hour hospital stay for Mrs. Loomis, and then she should get back to her life and I could move on to other patients.

Not until the next day did we finally meet face-to-face.

3

MEETING MELISSA

Melissa checked into the hospital. She had received three different oral antibiotics before I met her on August 14, 2015, almost three weeks after the bite. Immediately she was started on a penicillin-type IV antibiotic, my usual choice for infections. I've had success with it nearly every time.

Melissa took it all in stride, but Neil didn't find out she was in the hospital until after work. Like many people, he's not able to check his cell phone during work. As he left at the end of his shift, he turned on his phone and saw a text from his wife:

> I was admitted to the hospital for IV antibiotics.

Worried, he hurried to the hospital and found his wife on the fifth floor, the orthopaedic ward. Several other members of her family were there as well, and they found out, like countless families before them, that there's not much to do while sitting around a hospital. Sometimes I think it takes the long vigil of waiting at a hospital to make families slow down for an hour or two and simply talk to each other.

The IV drips proceeded, and they could watch that; it was interesting for about two minutes. They joked around, kidded Melissa about her lack of success as a wild animal tamer, and heard the only real item of information: that a Dr. Seth would be checking in that afternoon. Still, it was just a raccoon bite—Melissa, her dogs, some silly raccoon. Nobody was too concerned.

Since she was a diabetic, however, nurses were monitoring not only her vitals but her blood sugars and whether they were staying within normal limits.

By Saturday afternoon Melissa had been on the antibiotics for twenty-four hours, so I could check on her and talk to the family. I remember it was a beautiful, sunny afternoon and a Saturday, so I'd been spending my time with my family. What I expected as I made my way to the hospital was a quick consultation.

I got to the fifth floor, checked in with the nurses' station, and caught up on a bit of paperwork before making my way toward room 5514. What I found there was a highly skeptical group.

"How are we doing, folks?" I asked in my breezy, here-comes-the-doc way as I bustled in. Straight ahead lay the patient, a pleasant-faced, smiling woman. To her right was her husband, Neil. To her left, David, her brother-in-law. These were bookend grizzly bears, large guys who could have played middle linebacker for the Cleveland Browns.

Michelle, the patient's sister, occupied a chair to the left. Dad was to the right. The patient's stepmother was in the corner of the room. In short, I found a patient who seemed to be enjoying her IV, with a grim supporting cast.

"Hello, everybody. I'm Dr. Seth," I said, smiling.

They nodded and mumbled, returning the greeting without too much enthusiasm.

"I hear what we have is an animal bite and a very stubborn infection."

Melissa confirmed those facts, and I asked her a few questions about the last couple of weeks of antibiotics.

She told me she'd developed a fever, and her right arm was becoming more swollen and painful. I gently felt the arm and saw she was right. Her entire forearm was red, with two small puncture wounds on the underside of the wrist—two holes the size of points made by an ink pen. While they were almost healed over, the red streaks moving up the arm were impossible to ignore.

I examined all this carefully before turning to the family. "The IV antibiotics aren't doing what they should," I said. "This redness shouldn't be here, and neither should the pain Melissa is feeling. I want to keep her here, and tomorrow morning I'd like to perform what we call an I and D: Irrigation and Drainage. That means opening up the right arm [I pointed to the place where I would make the incision], and then we'll drain the pus." I looked around at the impassive faces. "Does anybody have any questions?"

They did. They had a lot of questions.

What was going on with a simple bite being this big a deal? Why had the doctors taken all this time with medicine that didn't work? And especially, could Melissa lose her arm because of this infection?

"Melissa will not lose her arm," I said definitively. "That's not going to happen because I'm going to clear the infection tomorrow morning."

"Are you sure?"

"I'll put it this way. I've never had an infection case when a patient ended up losing the arm." I wasn't blowing smoke; I'd never had that happen. However, as you will see, I'd never had a case like Melissa's, and I would live to regret these words.

One said, "No offense, Dr. Seth, but are you the best doctor for Melissa? She deserves the very best, and we just don't know you. Why are you so sure you can take care of this infection?"

Another asked, "Why should we use this smaller hospital, Dr. Seth? Wouldn't it be smarter to go to Cleveland, or somewhere else, to a larger one with more experienced people?"

Normal questions from a concerned family. I said, "Folks, I

understand these questions, I really do. You're right to make sure Melissa gets the very best care available. I want you to know there's no need to go somewhere else. Because you're not going to find a doctor anywhere who will take better care of Melissa than I will—smaller hospital or not. I promise you."

They looked at me a bit more intently. Michelle asked, "Are you the best surgeon for this kind of procedure?"

That's a common question. Who doesn't want the best? When someone you love has a serious medical issue, you'll do anything you can, ask any question you can, to make sure your loved one receives tip-top care.

But how can a doctor answer that question? Who can say who's *the best* for any particular situation? Every case is different, and every patient is unique. It's not credible to say, "Yes, I'm the best there is," and it's not reassuring to say, "No, there are bound to be better doctors."

In my town there are two hand surgeons who are just as skilled as I am. Either would have done a terrific job in this case or any other. I'd recommend them in a heartbeat. But it's futile to start making comparisons and ask who's the best. What's important is to be the best you *can* be, and to reassure the patient and the family that you'll do so.

"I can't tell you whether I'm the best," I said with a smile. "But I can tell you I will *try* my best."

I still didn't feel that I held the confidence of that room. Doctors face worried families all the time. You tell them what you can, you encourage them, and you do all you can to work at your peak potential. But on this day there was something else in the air—something nudging my subconscious, you might say.

I had a strong intuition about this patient and her family. It was as if I could hear a voice saying, *I want* you *to take care of this one. Take very special care of this lady.* In many cases like this one, a doctor might nod his head and say, "I understand. Let me give you the names of some good doctors in Cleveland," and then move on to another patient. But something inside was telling me not to let that happen. *I* was the one who was supposed to care for Melissa Loomis. I didn't know why, but the impulse was there.

Yet this family didn't seem to be sharing that impulse. They looked about as tense and ready to take action as the raccoon must have, at the last moment of clinging to that fence. For now, they were sitting on the proverbial fence, but they were very close to acting on impulse, and that impulse was to take Melissa north up Interstate 77 to the Cleveland Clinic. I could see that.

Their decision entirely—of course it was. But I was not supposed to lose this patient. I just knew that somehow, and I said, "No one is going to give Melissa better care than I will."

It was David who said, "You're our man, Dr. Seth. Just do everything you possibly can to help my sister-in-law."

"I give you my solemn promise she will not lose her arm." There it was—I'd made a promise. Now all I had to do was keep it.

Back at the nurses' station, I called the operating room to book Melissa's surgery for the next morning. The OR scheduler asked, "How long will you need?"

"Thirty minutes." That's standard for irrigating and draining an infected arm or hand. Open the site, drain, wash, and leave it open to heal. Most of the time, those unwelcome bacteria bugs clear out, and we're home free. What I didn't know was that these bacteria bugs were putting on the wildest party bugs can throw inside an arm. Imagine the craziest college frat house ever, with hundreds of kids fighting over one keg.

That's what was happening with the bacteria in Melissa's infected arm. Bacteria are not evil, by the way. They perform all kinds of necessary functions within bodies. In the case of infection, however, they go bad, and if we can't get them under control, things go badly for us.

The next morning I'd find out how badly.

4

FULL-SCALE INVASION

Church begins at nine. And I wasn't going to miss it, even with a morning surgery.

I told my wife I'd be operating at 8:00 a.m., finished in thirty minutes, and walking into church in time for the opening song. As a doctor, you learn how to fit normal family activities together with medical work like pieces of a puzzle. Otherwise your life is a big box of disconnected pieces.

At the hospital I greeted Melissa's family in the pre-op holding unit. I explained what I'd be doing and told them the surgery shouldn't take more than thirty minutes. Melissa, of course, was smiling. She was taken to OR12 and put to sleep as I left to put on my scrubs.

I was vaguely aware Melissa was a special patient, but that didn't mean I was intimidated by the surgery. This was bread and butter for me: in this case, a small raccoon bite with a stubborn bacterial infection. A three-to-five centimeter incision would be the only opening I needed to get the job done. I'd take cultures, leave the wound open for draining, and we'd be done.

As I changed, I was thinking about football—the Browns' NFL opener, a game in Pittsburgh against the Steelers. I love autumn and football season, and this, of course, was the year my Browns would make it to the Super Bowl. I knew it in my heart with the faith of a child. (The Browns, by the way, finished 3–13 that year.)

I put on my gown, pushed my fingers into the gloves, and greeted my surgical assistant, Trisha Carr, and my tech, Lori Fernandez, as well as circulating nurse Rhonda Hawkins. Trisha and Lori were there to help me with the surgery. Rhonda would make sure we had every tool and resource we needed. She'd also chart the patient's condition.

On the film screen of your mind, I'm sure you're staging this scene the way your TV shows have carefully trained you. You see the blinding lights on the operating table with a pool of darkness surrounding it; you hear the dramatic beeping of the patient's heart monitor; and you see lots of close-up shots of the compassionate yet heroic surgeon, played by a handsome Hollywood star (I hope) earning Emmys with every evocative eye movement.

Nope. It's only like that on network television.

It's a workplace, to be honest, with ordinary light and music playing, doctor's choice. That beeping monitor would drive us crazy! Don't you want us to be relaxed and comfortable while we're doing your brain surgery or heart transplant? We chat about our day as we work, just as others do at a bank or accountant's office.

That was the scene at 8:15 a.m., when we began.

Pen in hand, I'd drawn a line around the site of both tooth punctures on Melissa's wrist. Lori handed me a standard #15 scalpel, and I made my incision. What happened next drove any thoughts of football from my mind.

Pus shot from the opening—an unbelievable amount of it.

It seemed the small, red area, where the teeth had punctured the wrist, had formed a huge abscess, where it now stored a deep well of pus.

I took a few cultures, then began to probe around the opening I'd made. It was important to make sure the entire abscess was removed. The next step was to wash the wound with a pulsatile lavage instrument. (Don't be afraid of those last three words—I'm simply talking about an expensive high-pressure squirt gun. If I'd had one of these as a kid, I would have dominated in neighborhood squirt-gun fights.)

The purpose of the lavage instrument was to thoroughly clean out Melissa's wound. Yet when I continued to probe, the pus kept coming and coming. How much of it could possibly be in there? It was emerging with such intensity that my three-centimeter incision was beginning to expand. So I cut a little farther—down to the hand and up toward the elbow—and the pus was everywhere I looked.

This was more than a little disturbing. Normally the infection is contained in the area where an animal bite has occurred. That's how it usually goes. But this infection had traveled. I grabbed the lavage and began washing the bottom of Melissa's forearm. Glancing up at the clock, I began to realize this thirty-minute, no-drama procedure was going to take a little more time.

I knew now this wasn't the simple infection we assumed. I asked for a special pair of scissors and began to dissect into the muscles and tendons of her forearm. As I lifted muscle after muscle, more pus was revealed. How in the world had this patient been pleasant and smiling—actually going to work? Her arm and her hand had been under hostile attack all this time.

This wasn't the bacterial "party" I had described earlier; this was a full-scale invasion.

I looked at my assistants, and I saw their wide eyes matched mine. None of us had seen an animal bite quite like this one—raccoon or otherwise. It's not as if we hadn't been involved in intensive procedures before, but with this one there had been no advance warning. There was so much beneath the surface.

And that's something you can say is true about every part of this story.

5

"LEAVE NOTHING UNEXPLAINED"

One more time, I probed both ends of the incision. Each time I placed a hemostat (a forceps or clamp) into the arm, a new geyser of pus would erupt. It would take some thinking to decide how to proceed with such an infection. And doctors would prefer to do that thinking *before* going onstage, rather than in the midst of the act.

With Melissa's arm cut open hand to elbow, every muscle was on display, and each one was surrounded by pus. This bacteria owned the patient's right arm.

I needed to wash her hand too. The incision was close enough to allow me to examine the tendons extending through the carpal tunnel, the passageway connecting the forearm to the hand. Several tendons pass through that tunnel. I made a small incision there and hoped to see exactly what I usually saw.

I didn't. I saw a tunnel flooded with pus.

Now I knew I had to open the entire hand and fingers. There was no way to be certain just how far this infection had reached, but we knew this pus had to be evacuated. I sighed and got to work. I irrigated, I drained, then I found more pus and did the same.

But now, at 9:00 a.m., a grim silence had set in. There was no more joking, no more commenting on the music or chatting about our day. The sight before us packed a punch. Lori, Trisha, and I knew we were facing a far more lethal enemy in this infection. I also knew I bore the responsibility of bringing that news to the people who cared most about the patient—as well as the patient herself.

"I'll have to stop the incisions," I told Trisha. "We can't have the entire arm open. We just have to be sure we've drained it all." If we didn't, the bacterial infection would still be in business. The smallest pocket of it would be a base that could quickly multiply, and Melissa would be right back where she started. At some point she could even lose her arm.

I looked at the long incision, wrist to elbow. I had to stop. Surely we had it all. I did one final wash of the forearm. I was ready to place a sterile dressing on her arm when I heard a new whisper in my ear.

What I heard was the voice of a mentor from years ago. He had always told me, "You're not finished until you prove to yourself you've done everything possible to solve the mystery you've discovered and that you haven't missed anything. Leave nothing unexplained."

That was the voice of Dr. Paul Cook, the hand surgeon from Ohio. He had drummed into me the principle that when a surgery begins, it never stops until it's done—or at least until all possible information about the problem has been extracted. We owe it to our patients to know all there is to know once we've opened their bodies.

It made sense. But what was left to investigate in Melissa's arm?

I turned the arm to the other side and looked at four scratches from the raccoon's paws. No redness. No telltale signs of infection. If the infection was so bad on the bottom of the arm, how did I know the top wasn't also involved? Did I need to open the area around these scratches? I really didn't want to. We wanted to avoid multiple incisions at nearly all costs. Both sides of Melissa's arm would be open. That wouldn't be a good situation, but Dr. Cook's voice was adamant. *Answer every question that has an answer. Leave nothing unexplained.*

I asked for the scalpel. Trisha and Lori looked surprised; then they glanced at the arm and understood. It had to be done. I made a new incision on the scratch side, the top of the arm, from just below the elbow all the way to the knuckles. It was a very long incision, but that was the area the raccoon claws had reached. And after all, two little tooth marks had spread some unidentified poison up to the elbow.

As I cut, pus shot out from the arm.

Are you kidding me? Then I thought, *Thanks, Dr. Cook.* It would have been a disaster to miss all this pus.

Yet there were questions we couldn't answer. Scratches alone couldn't do this on the top of her forearm—could they? I shook my head and reluctantly began a new dissection. Infection was everywhere, and I took new cultures.

I told myself that as bad as this was, having to dissect both sides, at least we'd know what type of bacteria we were dealing with since we took cultures. Then we could choose the right antibiotic and destroy the horrendous bacteria. At least this is what I thought at the time.

The clock showed 9:45, an even hour and a half since I'd made the first cut. That's a long time for a bacteria hunt.

Now turning to the upper arm above the elbow, I held my breath and made one more small incision. No pus. I breathed out. I think my assistants did too. We'd reached the limit of the bacterial invasion. I washed her arm, top and bottom, one more time and wrapped it all in a sterile dressing. None of us said a word. The others knew exactly what I was thinking.

Do I really believe I can save this arm?

6
FACING THE FAMILY

I removed my surgical gown and gloves, thanked everyone for their help, and headed for those two doors that mark the entrance to the real world, where real families and their emotions lie in wait. I was apprehensive because I had neglected to update Melissa's family during the surgery. A sinking feeling came over me now that I was heading out to see them.

I'd told Neil and the others to expect me in thirty minutes. Then I'd gotten so absorbed by the terrible nature of Melissa's infection that I'd had no chance whatsoever to send word about what was going on.

And this, twelve hours after giving them my solemn assurance they could trust me. They'd been out here in the waiting room expecting a quick answer. An hour and a half later it was on the way, and it was a far worse answer than they anticipated.

I racked my brain for the right words, but what could I say? I'd promised them this would be a simple and straightforward procedure—just evacuating a small amount of bacteria, the work of a few minutes. And now I had to inform them that we had a grave, possibly life-threatening situation in front of us.

The five who made up Melissa's group in the waiting room stared me down as I came through the doors. Being Sunday, things were slow at the hospital. No other families were around. I sat and met their eyes. It was clear they knew, after all this, that something wasn't right.

"I've dealt with hundreds of infected bites," I said. "And I've never seen so much bacteria in the form of pus. Melissa's arm just wouldn't stop draining. We did everything we could to clean out her arm, and she's resting comfortably in the post-anesthesia care unit."

I took a breath and continued, "I took about six cultures so we'd have what we need to identify this bacteria. There are a great many varieties in the world. With the amount of pus I saw and drained, I'm sure something will grow out in the microbiology lab, and then we can put a name on it. That will tell us what kind of weapon to use in fighting the infection."

The family took in my words quietly, trying to deal with the fact that Melissa had a much more serious fight on her hands than they'd been led to believe.

"We all saw the redness and swelling," I said. "But however you imagine the condition of her arm, just magnify that by ten—it's far worse on the inside than we could have suspected. Both sides of her arm, top and bottom, elbow to fingertip, had to be completely opened." I showed them what I meant on my own arm as I spoke.

I hesitated and added, "If you're able to handle it, you can see pictures of all this."

They all nodded, some saying, "Yes, please," and "We'd like to see them."

I pulled out my phone and began to scroll through my large group of pictures of Melissa's arm. The photographs said what my words couldn't. I saw the family cringe as the images were revealed, and I wondered how the patient herself would react. I'd only been with her for a total of twenty minutes the afternoon before. I still didn't know her well. In forty-eight hours the wound dressings would be taken off, and she'd be confronted with something shocking.

Better than any doctor, these loving family members could prepare her for what she'd see.

"We're fighting back," I said. "The infection fooled us with its size and its spread, but that doesn't mean we can't handle it. We're going to put Melissa on the strongest antibiotics we have, and I'll also be spending time with the infectious disease doctor—this is his field. In that area, *he's* the best of the best." I'd be sending him the same surgery pictures so he could see the size of the monster bacteria we were dealing with.

David often spoke for the group. He asked the question I knew was coming. "Doc, yesterday you said you could save her arm. I remember you told us you'd never had to amputate an arm before in this kind of situation. How do you feel about that now?"

My heart sank, but these people deserved a good answer to their question. What could I say?

7

DOCTOR AND PATIENT

Twelve hours earlier, I'd been so sure of myself. After doing something hundreds of times, we start to think we've seen all there is to see, and that we know all there is to know. But I'd been taken by surprise, and I couldn't take back my bold promises. You can't unring a bell once it's been rung.

I could have doubled down on my promise: *She's still not going to lose that arm! I'll come back and wash it again in a few days, and there will be no bacteria by that time. Everything will heal, folks.*

But I knew that wasn't the truth. The *truth* was this was the most challenging case I'd ever been presented with since I'd started in orthopaedics—not in relation to technical skill but because of the still-unknown but powerful ravaging bacteria confronting us.

I looked at David and gave him a sad-but-honest answer to his question about saving his sister-in-law's arm. I said, "I don't know."

I paused, letting them take in those words, then continued, "I don't know if she's going to lose her arm now. But I can tell you one thing: I will work day and night and do everything I possibly can do to see that she keeps it."

I think they all had some idea of the seriousness of the situation. What we didn't discuss was the possibility that the infection might continue its journey from Melissa's wrist to her shoulder and points beyond. I'd made it clear that now the bacteria had an outlet, a place to disembark.

They asked more questions, and I fielded them. We all shook hands, and I hugged Michelle—a worried sister—and told her everything was going to be okay.

Doctors work to heal the body, but a surprising share of our work involves nurturing the spirit. We do all we can to reassure people everything will come out all right, but we always walk that tight line between offering comfort and avoiding harsh reality. The best practitioners find ways to provide an honest prognosis with genuine hope and encouragement. And we hope people understand we're not God; there are situations we simply cannot master, battles against infections and diseases we cannot win.

I hoped and prayed this occasion was not one of those.

I walked out of the room and down the hallway, into the post-anesthesia care unit. I looked at the clock. By now my family was out of church. They'd surely watched the door, expecting me to walk in at any moment, so they could gesture and show me where they were sitting. I hadn't made it, but they'd understand. They knew that in the world of surgery, all other bets are off.

To my surprise, Melissa was already awake and resting comfortably. I walked up to her bed and said, as gently as possible, "Melissa, we operated on you for about ninety minutes. Longer than we initially expected because there was more infection there than we thought. It's a challenging situation, but you're going to be all right."

I didn't yet realize what kind of patient Melissa was. You don't have to break things to her gently; she has an inner constitution of iron. As I finished saying my piece, Melissa replied, "I hope you got all that raccoon bacteria out of my arm."

"I looked and I looked, Melissa, and I can tell you I couldn't find *any* raccoon left in that arm. I beat every single one of those bacteria out of there."

She smiled, and I was pleased with the conversation. This wasn't the best time to give her a foreboding preview of what was next. She had twenty-four hours to prepare for the current condition of her arm. Her family would be the ones to talk with her first because they knew her best. I could fill in the details afterward.

Soon I'd changed back into my street clothes and was heading to my car. It was 11:30 a.m. I dialed my wife. "What happened?" she asked.

"Long story," I said. "Short version: I have an arm to save. It's in terrible condition, but I've got to save it."

8
THE DOG LOVER

It just had to begin with dogs. If you knew Melissa, you'd take that for granted because dogs are her abiding passion. She can't get enough of them.

Now, she'll insist that she's not one of those really intense animal people—the crazy cat lady or the fellow whose whole life is wrapped around competing in the National Dog Show. She used to own cats. When she discovered dogs, however, some tiny gear inside her clicked into place. Dogs—and finding them homes.

Some people are born for fishing or golfing or European travel. Melissa was born to play matchmaker between the worlds of people and four-legged, tail-wagging animals.

She had signed up as a volunteer with the local dog warden. It sounded like fun—something to do about ten hours each week: walking the dogs, playing with them, and helping them go home with a loving family.

It wasn't long before she realized these were the most satisfying hours of her week. Melissa had always loved people, but dogs? For her they were special—the intensity with which they loved, the absoluteness with which they trusted, the size of their devotion. Nothing gave her a better feeling than seeing the face of a new owner taking home an ecstatic pooch.

She would come to the shelter, see all the dogs craving a little human affection, and dream about finding a wonderful, loving home for every single one of them, maybe with boys and girls to run alongside through the outdoors, romping with their new best friends. The people's lives would be a little richer, the animals would think they were in heaven, and the world would be—ever so slightly—a better place. World peace one pooch at a time.

Her passion for the cause was noted, and soon she was a board member of Friends of Stark Pound, whose motto is "We Bark in Stark!" It was a fund-raising operation that existed to put rescue dogs and families together.

I get it; I like dogs too. But wow! I'm always intrigued by what makes people do what they do and love what they love. One day I asked Melissa, "With all the ways you could choose to spend your spare time, all the countless good deeds there are to do in this world—why dog rescue?"

She simply offered her Mona Lisa smile and said, "I want to make the world a better place by helping people and dogs interact."

Something else. I think Melissa senses that, like nothing else, a dog can reach something deep inside a broken spirit to begin the healing process. As a doctor, I can appreciate that. This is why Melissa's ultimate dream involves putting dogs and military veterans together. She'd love to become involved with an organization such as Wags 4 Warriors (you've got to love the doggy language of all these groups). W4W is based in Ohio, helping veterans who struggle with post-traumatic stress disorder or traumatic brain injury, providing them with loving and helpful dogs at no cost.

If you're listening, W4W, take my word for it: you'll never find a more passionate champion for your cause. No one but an actual dog would be more devoted.

Melissa loves having dogs around. Working at the shelter, she ended up bringing home a couple of them she particularly liked. If it were up to her, she says, there would be at least ten pooches of all kinds roaming the Loomis household and exploring the backyard in all its wonder. But Neil does have his limits. Vincent and Vivian will have to suffice for now, as well as another fifty or so canines she's caring for at the rescue center. All of them are her children.

Her day begins, as it did on that fateful morning, with the dogs serving as a gentle but reliable alarm clock. She drives to her job at the Canton Regency, an assisted-living facility, where she works as the director of food services. I'm told she's as well loved there as I'd expect.

Fifteen years ago, Melissa discovered she was a diabetic. For half that time she's been dependent on an insulin pump to help regulate her blood sugar.

Melissa's character is essential to the mechanics of her destiny, just as yours is to your own future. I find it difficult to get across in words. I'm sure we all know two or three people who are "salt of the earth" types: just plain good folks without a mean bone in their bodies, who do about ten times their share of good in the world, then show uncommon courage when the world returns the favor with evil. Melissa is one of those good folks, but this story, as far as she's concerned, turns on something rarer and more distinct than that.

The writer G. K. Chesterton said, "Angels can fly because they can take themselves lightly." That's close to hitting the mark in this case. Melissa calmly lived out this amazing period of life in which she ventured through ups and downs that would have exhausted the spirits of most people. Never for a moment did she shake her fist at the heavens or question my recommendations for her care. She loved to say she was "just along for the ride."

The last time she told me that, I said, "Yeah, right. It looks to me like you're the one driving the car, and I'm over here sitting in the passenger seat."

Except there would also have to be a couple of dogs in the back seat, with their heads out the window, jowls flapping in the breeze. One way or another, this was going to be the craziest ride of our lives.

9

GETTING THE PICTURE

On Monday morning my word was *focus*. I needed to shift Melissa's plight to the back of my mind to be fully focused on my other work. Normally I operated at a different hospital, so Melissa was across town from where I'd be.

I was grateful when the morning cases went smoothly. By noon I was caught up and had about an hour before patient appointments in my office would begin. That hour would afford me forty-five minutes to visit with Melissa at the other hospital. By now her family would have spoken to her, and she'd have a better understanding of what we'd found.

I had no idea how she was going to react. I knew her as a courteous, friendly patient. I'd talked with her for a few moments post-op, but she was heavily medicated at the time. I wanted to be truthful yet compassionate—that difficult but healthy balance. I was going to let a nice woman know that two small tooth punctures could mean, well, it could mean the one eventuality I had hoped to avoid: the loss of an arm.

I stopped at the nurses' station to look at Melissa's charts and check the status of her recovery. I also had to consult with the infectious disease doctor about antibiotics going forward, so I paged Dr. Badie Al

Nemr and got a quick callback. Dr. Al Nemr and I have a long history of working together on several infectious disease cases.

There are more than one hundred different antibiotics, though they all are variations on a few basic drugs. If you've ever had a sinus infection, you've probably taken an oral antibiotic. But there are also intravenous antibiotics. In most cases my patients are on intravenous antibiotics for no more than forty-eight hours while in the hospital, then take them orally for another week at home. I can't remember the last time a patient had to return for a second course intravenously.

Whenever I tell this story, someone naturally asks about the raccoon and rabies. But Melissa had received a vaccine for that immediately when she got to the ER on the day of the bite. So I didn't believe rabies to be an issue. After giving Dr. Al Nemr an update on Melissa's surgery from the previous day, I told him I'd send photos, via our secure hospital line, so he could get an idea of the infection.

I was glad I'd taken all those pictures before the wound dressings were applied. A picture, as we all know, is worth a thousand words, and that holds true with consulting doctors. We also wanted to avoid redressing Melissa's arm multiple times because doing that would put her in a lot of pain.

Fewer than sixty seconds after I sent the photos, my phone rang. I picked up, and the infectious disease doctor said, "That's the worst infection I've ever seen, Ajay. I mean that."

"That makes two of us. The good news is, I drained so much pus that we should have some kind of bacteria growth from our cultures. That should tell you what you need to know."

One or maybe two antibiotics can be used to kill a given infection. Since this one was so severe, Dr. Al Nemr decided to use a more powerful antibiotic, though safe, of course, and to begin a second one. This way we could hold off the infection until we had the lab results from the cultures, then proceed accordingly.

These were strong medicines, but for me, it wasn't just about ridding Melissa's body of the bacteria. I was concerned about saving the muscles in her arm that had been damaged from the infection.

We finished our call, and I walked down the corridor to Melissa's room. I thought again about how to bring this news to her. And, of course, to do so with compassion and warmth. The best approach was to sit down beside her bed and simply explain the facts and tell her what possibilities lay ahead.

10
OMINOUS QUESTIONS

Neil was with her when I arrived. Melissa was sitting up in bed—again not a surprise. Her entire arm was wrapped with a soft dressing and covered with an Ace bandage wrap.

All of it looked neat and tidy, but Neil and I knew that beneath the bandages her arm lay opened up from top to bottom, on both sides, and every single finger on her hand was open as well. Her husband, sister, and the others would have told her a little bit about this by now, but she hadn't seen the pictures—or the real thing. And sometimes you have to see it to believe it.

I greeted them both, asked Melissa how she was feeling (just fine, of course), then took a seat near the foot of the bed. They both waited quietly to hear my news.

"Melissa, the infection is much worse than I ever imagined. I'm sure Neil has brought you up to speed on that. I've taken care of hundreds of these bacterial infections through the years, and I simply haven't seen anything close to this. The infectious disease doctor has seen the photos, and he hasn't seen anything like it either. I know all this is a little difficult to imagine, since you're looking at a neatly wrapped arm at the moment. But I want you to be prepared for what you'll have to see tomorrow."

"I understand."

"Now, to get this infection out of your body once and for all, I had to make an incision on the bottom of your arm." I pointed toward the place, and continued, "When I did this, pus began to come out—actually, it began to *erupt*, there was so much of it. As powerful as that pressure was, the pus wanting to escape, I can't imagine how you continued to function. And you were at work, doing your ordinary chores, right?"

"Yes, I was."

"Well, the infection moved down into your wrist and upward all the way to your elbow. You must have a far greater pain threshold than I—or most other people—do. Since there was pus everywhere I checked, I had no choice but to keep making incisions, and I couldn't close them.

"If I had, the remaining bacteria would keep reproducing itself—it would be doing that right now—and the infection would grow only stronger and stay on the move. We always want to close an incision. The idea of keeping it open, with your arm, is to encourage that bacteria to go somewhere else."

She and Neil were nodding, clearly following what I was saying. But Melissa was so calm, I might have been talking to her about painting her house.

"I want to do a dressing change in twenty-four hours. I'm going to give you a medication that will help you relax, and then we'll change the dressing. At that point you'll see your arm as it is now, in a way you never would have imagined. We'll also find out about your fingers—that is, what kind of function you have with them. I hope the muscles that move them are still viable."

That last part, of course, was the gut punch. It suggested the possibility of losing the use of her right hand.

She looked once at her husband, smiled back at me, and said, "Okay, Dr. Seth. We gotta do what we gotta do to save my arm. We'll see you tomorrow, and I guess we'll unwrap my arm and go from there."

Had she not taken in the words that had just come from my mouth?

I'd described an arm that was basically a battlefield after the bloodshed. Most people couldn't handle seeing it, hearing about it, certainly not *having* it. This woman was either unnaturally calm and powerful on the inside or outright crazy.

But already I knew better. This was option A, not option B. If a doctor had just briefed me like this about my own arm, I'd have needed to be sedated. They would have found it necessary to move me to a soundproof hospital room. But Melissa was strong, and maybe something more—maybe that something was *chosen*. I'll leave that for you to decide.

As I rose and prepared to leave, almost shaken by how well this had gone, a small object caught my eye—something completely out of place. I looked and saw the last thing in the world I should be seeing in her room.

I saw a stuffed animal—a *raccoon*?

Three things sat on Melissa's window ledge: her cell phone cord, a small vase with a flower, and this stuffed animal.

What's wrong with this picture? Who in the world would bring a plush, cuddly symbol of the creature that attacked and terribly endangered Melissa?

"Melissa, who brought you the stuffed raccoon?"

She laughed and said, "Isn't that cute? One of my friends thought I'd like it."

I thought I had crazy friends—the kinds of friends who throw slightly tasteless pranks at you just to get a reaction. But none of them would do anything like this. *Unless* they knew their friend extremely well, enough to know she could take it, like Melissa, so comfortable in her own skin that she would take nothing the wrong way.

I looked at Melissa again—seeing her for the first time as she really was. I said, "Well, we've all got some crazy friends, don't we?"

She smiled in her genuine way and said, "See you tomorrow."

Looking back, I realize that's the moment when we truly bonded—over that darn stuffed animal. I walked down the hall and thought, *I think she is going to be all right.*

PART 2
THE INFECTION

11
PUSHING FORWARD

Our parents would have given us the moon if it were possible. Instead, they gave us America.

My mother and father immigrated to the United States for the benefit of their two children—my sister Angela and me. It wasn't out of necessity. In India they were comfortable, living in relative luxury.

They gave up all that and traveled to America because of the possibilities their children would find. They came here and settled on the floor of a YMCA. They believed in the two of us, that we would make something of the American Dream. They taught us every day that *opportunity* is the most wonderful idea imaginable. And America is the land of opportunity, a place to choose and pursue our own dreams rather than have someone choose them for us.

Angela and I understood the sacrifices our parents had made so that our horizons would have no limits. Starting out as a child with that perspective changes things.

I was free to "choose my own adventure," as in those children's books. When I received a local award as a teenager, a newspaper reporter asked me about my career plans. I didn't hesitate with my answer: I wanted to be an orthopaedic surgeon. That was my story, and I was sticking to it. I still am.

That story was thirty-two years in the making. However, my public education, undergraduate and medical school, research year, and residency required that much time—basically my youth. My parents had also taught us that good things require time and sacrifice.

My final year of formal education, at Allegheny General Hospital in Pittsburgh, Pennsylvania, was called a *fellowship*. That's where I began to learn about specialized surgery in the upper extremity: shoulder, arm, elbow, forearm, wrist, and hand. This became my field of emphasis.

As I attended my last graduation, it was good to have all the books, examinations, and certifications behind me. The real world and a real career beckoned to me. It also worried me! I'd be in a practice, but I'd have my own patients, my own surgeries, without some wiser, more experienced physician looking over my shoulder. I'd be immersed in the world of bones and joints, tendons and muscles, skin and soft tissue coverage, blood vessels, and nerves.

At the graduation ceremony I heard a speech that captured my imagination and has never let me go. One of the attending hand surgeons, who taught me a great deal, used only a few words; five minutes and his speech was done. But I've never forgotten it.

He was speaking to the three of us who were "outgoing fellows." (I hope my friends *still* find me to be an "outgoing fellow.") He said, "You are responsible, as an orthopaedic surgeon, to push the science forward. You're not here to tread water but to swim into deeper seas. That always involves the risk of getting in over your head.

"But if you don't do that, if you decide to play it safe and stay in the shallow water—that is, do your job, cash your checks, and go home—then you've failed to use your God-given talents for their true purpose. Science is a journey. Our generation must push further than the one before it. And if you don't, who will?"

Please realize that I didn't feel ready to perform even a common carpal tunnel surgery by myself. Nerve repair? I lacked the nerve myself!

I was a rookie, no matter how many diplomas hung on my wall.

Yet he was challenging me to perform some kind of groundbreaking surgery in the future.

Most of us keep our heads down and struggle with the headaches of the hour. Gazing into the future, boldly going on missions where no one has gone, is a little scary. I know it was for me, a young man on his graduation day.

But thinking about that has a way of recalibrating your compass. You realize you're not simply walking the next mile for the goal of dinner, a nicer vacation, and early retirement; you're on a journey, and it's bigger than you are. When I retire, I want to be able to say I gave something back. That's another lesson I learned from my parents.

I understood, through the perspective my parents had given me, that science, too, was a land of opportunity. My parents had brought me here, my educators had taught me here, and the rest was up to me. It was all about finding the right place to make a mark and help build a future for others. The theme of any life, especially a doctor's, should be service.

There was a simpler way of touching the future. Before that fellowship I began building a family.

Kim, my wife, is a physician in a family practice. We'd gotten married six years earlier, in 1998, and we have two children. Jaideep is our teenage son, and Trinity, our daughter, is a preteen. Then there are two dogs, though ours haven't taken on any dangerous raccoons lately—if they do, I'll just watch from the window, thank you very much.

My period of practice has flown by, and, of course, I've grown in confidence and skill with each year. But some part of me has always kept watch for that moment described by my mentor, that opportunity to make a significant contribution to the science itself, to pay my debt to the future.

Not that I made that connection when the moment came. The day I sat at a football practice, answered my phone, and heard about a patient with a raccoon bite, the case seemed as ordinary as all the others.

A forty-three-year-old woman with a couple of tooth marks? In the journalism world, the old cliché is that "Dog Bites Man" is not an important news story—too mundane and ordinary. In our world, "Raccoon Bites Woman" is similarly no reason to alert the media.

At the beginning of this case, I did have some twinge of doctor's intuition about Melissa Loomis. *You need to take special care of this one*, that voice whispered. But it wasn't something I gave extra attention. The impulse was just that—a vague instinct not to let this patient out of my hands.

After her first surgery I knew that I was looking at more than a "Raccoon Bites Woman" story. "Monster Bacteria Rampages Through Woman's Arm" was more like it, and the truth was, it was the worst infection I've ever seen. It might be worth alerting the *medical* media—the journals, the experts. But nothing about it said, "This is a game changer."

It would be my toughest, most challenging case, mentally and physically. It would involve a remarkable patient. But I wasn't close to putting together just how significant this case was and what was truly at stake.

Destiny had used a lot of moving parts to make it all come together for Melissa and for me. It was beginning to appear that each aspect had been *chosen*—which is when the idea of higher purposes started to become evident.

12

ON THE OTHER HAND

The night after meeting with Melissa and Neil and seeing that crazy stuffed animal, I began to worry about Melissa. How would she respond emotionally when she saw what I had done to her arm?

It's one thing to be told about it. It's another to see it, that collection of incisions still open—inches from your eyes and attached to your own body.

More importantly, I worried about her fingers and their range of motion. What if she'd already lost the use of them? Was she really prepared for what that could mean?

What I wasn't worried about was the infection. We'd irrigated, we'd drained, and now we were hitting it with a double shot of powerful antibiotics, administered through IV. I hadn't seen an infection yet that would stand up to such an antibacterial barrage. The issue was how much damage the monster had already done—whether it had left its victim a usable arm.

On Tuesday morning I made sure I was out of bed early so I'd have plenty of time to see and evaluate Melissa. If she could move her fingers completely, or if she could at least partially close her hand, I'd know the tendons and nerves were working. I was anxious to find out, and because I had no other patients in that hospital, I could spend all my time there with her.

I stopped by the nursing station on the fifth floor to ask for some help with supplies so we could unwrap Melissa's arm. I discovered they were already giving my patient wonderful care—half because they're great nurses, and half because Melissa's an uncommonly calm and optimistic patient. They'd bonded with her already. One of them said, "Please tell us she's not going to lose her arm."

"The surgery went very well," I said. "I'm sure her arm is going to be fine."

I found Melissa sitting up in bed eating breakfast, with Neil by her side. She seemed glad to see me. Her shirt made an impression on me: blue with all kinds of emojis. I hoped she would always reflect the smiling and laughing ones, rather than the angry or weeping emojis.

"Today we're removing your bandages," I said. "You'll see your arm, and we'll see how much function you have in that hand."

"I'm more than ready to get this thing off."

I took a pair of scissors and began slowly snipping through the bandages on her arm. I'd applied a thick dressing. I hadn't seen the wound since the operation forty-eight hours earlier, so I honestly didn't know what to expect. I was curious about not only her nerve function but also whether the muscles were working.

I looked up and saw the eagerness in the eyes of the patient and her husband. This was what they call on reality TV shows "the reveal." But it wasn't a quick *voila!* moment. It took some slow cutting, which of course raised the level of suspense. I finally made it to the last layer, underneath which Melissa's ravaged arm awaited our eyes.

I said that little, almost subconscious prayer we all say; I know all three of us did. *Please let everything be fine.* Then I pulled back that last bandage.

I can still hear the gasp. I didn't say a word for at least a minute. I wanted to give them an opportunity to register what their eyes saw. That's important.

From my perspective the arm didn't look bad at all. But I knew their perspective would be very different. It would be shocking. I'd explained what I did in surgery so they could make sense out of what they saw and begin to think realistically about what came next.

"Doc," said Melissa. "Are you sure you can save my arm?"

It was the question on everybody's mind, from the nurses' station to my own.

"Absolutely," I said without skipping a beat. "I'm 100 percent sure that I can save this arm. Remember—I've never amputated an arm because of an infection. I don't see why I should begin now."

I could see them trying to cling to those words even as they looked at the sad vision of an arm devastated by bacterial attack. I wondered if they were thinking the same thoughts I would have: *This doctor is blowing smoke. It's time for him to stop talking about his skill and start showing us some of it!*

I honestly don't think they saw it that way. They'd heard me out, and they seemed all in. For better or for worse, they believed I was there to do everything in my power to save Melissa's arm. Otherwise, I imagine they would have asked for a change of doctors and facilities.

I put away my scissors and sank into a chair next to Neil. Melissa said, "It's a good thing I'm left-handed, isn't it?"

That comment took me by surprise. How had I not asked that question myself? It's of supreme importance when one of your two arms is at high risk. In the general population, nine out of ten people are right-handed. Nine out of ten with infected right arms would be in danger of losing the hand they did nearly everything with—their key tool for interacting with the physical world.

Once again Melissa defied the odds. Her left-handedness was a small but significant piece of good news. We all want two good hands, but if we had to lose one, we wouldn't want it to be the dominant hand. Melissa was right: this could have been worse.

"That's a great way of looking at it," I said. "Do you have any questions?"

"Not really."

I expected a bombardment of them. Offhand, I could think of about ten big ones she might have asked. But she couldn't think of a single question.

"Neil, what about you?"

"Oh, heck yeah. I've got questions." And he did have a number of them—some I could answer, and some I couldn't. It all came down to his final one: "So you're really sure you can save this arm? Because the more I look at it, the more worried I am."

"I get that. But, I promise, I'll let you know when to worry. You two sit back—I've got this."

13
CAN THIS ARM BE SAVED?

I sat at the computer and began clicking the mouse. I was eager to get the results on the cultures we'd taken during the surgery. I had to navigate through about fifteen screens to get to the tab I was looking for, the one marked *Results*.

I was aghast. Every bit as much as I'd been during the surgery itself. The words read,

NO ORGANISMS SEEN

Nothing.

Basically that was impossible. I had seen the greatest proliferation of bacteria in one person ever in my career. No organisms detected? Nothing to see here? Please move on?

You might as well have gone to New Orleans after Katrina and told them there was no evidence of wind or water.

It was more than just a curious thing. We needed that information so we'd know exactly what we were up against. Without the identity of the perpetrator, we wouldn't be able to choose the right weapon. We'd be fighting while blindfolded, to some extent at least. One antibiotic isn't the same as the next one. It's highly critical that we make a wise selection.

I called Dr. Al Nemr and passed on the information. "All those cultures we took," I said, "they tell us nothing. No bacterial organisms."

"That makes no sense," he said.

"So the question is, where do we go from here?"

He paused and said, "I keep thinking of those pictures you showed me. That infection is for real. I'd say we put her on the strongest antibiotics we have."

"I agree."

During the last round, we'd put her on stronger ones, but now we would try the most powerful ones available. There are reasons they should be a last resort, but we were nearly out of options. I headed down the corridor for Melissa's room. What was I supposed to say this time? It would be nice not to begin a meeting with, "You won't believe this, but . . ." Nothing about this case was following the script.

"You won't believe this," I began, "but there's no trace of bacterial species from the cultures we took. I was absolutely convinced we'd identify the bacteria. That's how we choose an antibiotic."

"Well, what should we do?"

"We'll put you on our very strongest medications. Surely they'll work this time—we just won't know exactly what we're fighting. We just need to do this quickly because the charts say you're running a fever."

Now it was time for what I hoped would be the good news we all needed.

"Time for the main event," I said. "I want to see you move your fingers." I looked over at her right hand.

Melissa began trying to wiggle her fingers. Nothing happened. I sat quietly—no pressure. Inside, I was urging, exhorting, pleading with those fingers to *move.*

I've seen this before, and I know it's human nature to become discouraged quickly. "I can't do it, Doc!" a patient will say. "I can't move them!" But Melissa expressed no panic and simply kept trying.

My deep concern was that the muscles in her forearm were no longer functioning—that the bacterial infection had damaged them. If all function was gone, we might be forced to amputate, and that was something I'd promised would never happen.

As I was beginning to give in to discouragement, I detected the smallest quiver in one of her fingers. "I did it!" Melissa said.

"Yes—that's great! It's a good sign. Melissa," I said calmly, "I want you to show me the muscles in your arm are still working."

I honestly believe she was calmer and more patient than I was. After a minute, her fingers began to curl in a bit. Inch by inch, or rather fraction of an inch by fraction of an inch, the fingers were starting to move downward. *Yes!* I wanted to celebrate, but I kept my cool. *We're going to save this arm.*

Melissa managed to bring her fingers about 30 percent of the way to her palm. I looked at her and said, "Considering what your forearm and wrist have been through, that's tremendous progress. This is a huge victory."

I looked up at Neil. I could tell he didn't particularly like looking at the wound, but a smile was threatening to break out across his face. His wife told him, "See? I told you—I'm going to be perfectly fine. I'll be out of here and back to my normal routine in no time."

I looked at her again, as I had when she talked about the stuffed animal, and realized what a unique, positive, and hopeful patient she was. If only every individual who walked through our doors could be like this one.

Most patients ask, "How long do I have to stay cooped up in here?" Melissa took charge of her fate or tried to. She decided she didn't plan to hang around. Not that I don't believe in the power of conventional care, but the human mind is the most powerful healer outside of God himself.

Melissa looked down at her right hand, willed her fingers back to their original position, and smiled with satisfaction.

I said, "It's going to take some work to get your arm back to normal. We're going to start occupational therapy. Also, we're adding yet another powerful antibiotic, and we definitely want to see that fever come down. And during the day, work on moving that hand as often as you can."

I also had to think about closing up her arm once we were sure the infection was gone. And I was pretty sure it was because after what I'd seen during surgery, this arm was beginning to look better. There were no signs of bacteria. Everything looked pink in a good way: healthy, healing.

A nurse and I rewrapped the arm. I already knew Melissa had a high pain threshold, but it was also possible she'd had pain meds before I arrived. "I'll check in with you tomorrow," I said. "For the next few days, just make yourself comfortable and let your body take in those antibiotics."

"Will do. See you tomorrow."

I walked out the door with a smile on my face. I liked working with this patient. I also told myself, "I'm going to honor my promise to save that arm."

14
CLOSING TIME

Melissa continued to run a fever of 101 over the next few days—even with all the antibiotics coursing through her system. Unfortunately, because of the ongoing fever, I had to take her to the OR and rewash the arm daily.

Each day the arm continued to show improvement, but her fever continued. On Saturday the wound looked good enough for me to tell her, "Tomorrow you're going back to the operating room, and we'll begin to close up that arm."

Her eyes lit up. "Does that mean I can leave the hospital afterward?"

"Slow down a little, Melissa! We'll need to keep you three to four days after that for recovery."

"Whatever you say."

For the closing-up process, I felt fairly certain we'd be using what we call a "wound VAC." The VAC stands for *vacuum-assisted closure*. Imagine a suction device, cellophane-wrapped over an open area of the body. As the wound VAC slowly removes fluid and decreases air pressure on the body, it brings the skin closer together and allows the wound to heal more naturally. Modern technology has given us some cool toys.

One of the best parts about the wound VAC is that you can use it at home instead of hanging around a hospital. The VAC requires a bit of maintenance, but I felt certain Melissa and Neil could handle changing and cleaning it at home.

Meanwhile, her fingers were finding a little more range of movement each day, and that was highly encouraging. Her forearm was making a comeback. I was beginning to think we could send her home two to three days after the Sunday surgery.

I work a lot of hours during the week, with surgeries at the hospital and patient consultations at the office. Sundays I'm at home, and I place my family as the focus of that time. I try not to discuss my work. Since Kim is also a doctor, we have to double our efforts to keep our home a medical jargon–free zone. We want our kids to have a real mom and dad, not a doctor and doctor.

For that reason, I wasn't talking about Melissa and her infection at home, other than to mention I'd be operating again on Sunday morning, the same as the week before. Hopefully the surgery would be a little calmer and more predictable this time. And I'd take another shot at walking through the church doors at straight-up 9:00 a.m.

Sunday morning I rose, showered, and drove to the hospital. The OR is on the second floor, and I hustled there because I was running a few minutes late. I have no clue why I tend to be a few minutes late, which is probably why I'm always a few minutes late.

The plan was to close the incisions. But first I would repeat the procedure from the week before because I wanted one more shot at getting some cultures. When Melissa's family went to the waiting room this time, I didn't offer a time frame. I'm sure they understood why. Last time I'd described a twenty-minute procedure, and it stretched to ninety minutes.

It was déjà vu as I stepped into the operating room: same day of the week, same patient, and the same circulating nurse and surgical tech. This was a good thing because it meant we had a team that was up to speed on this unique situation. We all were excited about closing those arm wounds after seeing how wide an opening I'd made in the arm.

After Melissa was asleep, we prepped and draped her right arm, and I carefully took my cultures. Once again I took the pulsatile lavage (the Super Soaker for doctors), washed out all areas, and took a good look at what was going on inside. I felt a new wave of encouragement to see that the muscles, nerves, and arteries looked reasonably good. I went over all the different areas with the culture swabs.

I did find two points of concern. Muscles should look beefy red, but I saw two light pink areas. I wasn't pleased about this at all. I selected a Bovie—an electrosurgical pencil that sends a small current to help with cutting and cauterizing structures—which can stimulate muscle. If a muscle has lost its function, it won't contract. When I touched one of those muscles, nothing happened.

I checked to ensure that the machine was turned on, and I verified that wasn't the problem. I touched the two areas again—nothing. The inescapable conclusion was that the patient had some muscular sections that were no longer functional; the good news was that there were only two, and they were relatively contained.

So what do we do with dead muscle? Remove it. Otherwise the bacteria will have an empty home to use as a base in starting up a whole new attack. So I very carefully, very tenderly removed those dead portions of the muscle. My thinking was that this would be my last major move on this case. It would settle the issue.

Then I began to close the incisions very loosely. A small area at the bottom of her wrist was still very swollen, so we couldn't close the entire incision. We applied a wound VAC, and we were done.

I offered a thumbs-up to the team. This had gone so much better than the appalling discoveries we'd made exactly a week ago. Melissa was easily awakened, and we moved her to the recovery room.

When I walked through those two automatic doors, I was much more upbeat this time. I smiled at the family waiting there. "Hello, folks," I said. "Things look good."

I gave them an explanation of our procedure, the two muscles I'd removed, and how the rest of the muscular areas were by all appearances beefy and healthy. I described the wound VAC and said, "We want to observe Melissa for two to three days, then release her to her own home and bed. We love having her here, but I get the idea she's a little homesick. We will teach all of you how to change a wound VAC. I think you can handle it just fine, but if you don't feel comfortable with that, we have people who can come to your home and change it for you."

They nodded and, in some cases, even smiled. I think we all felt we were beginning to see the light at the end of the tunnel. If the idea of leaving the hospital isn't a moment to celebrate, then what is?

Monday would be all about other patients; that seemed pretty likely. After all, Melissa was simply recovering from surgery and letting her wounds begin to close up. There was no reason to expect any big news about her situation.

15
ICU

Why do we always expect life to be predictable? We make our plans, we organize the world, then real life seems to love tangling up the whole thing.

I had planned to check on Melissa around lunchtime, after my usual Monday morning operations. But I didn't make it to lunchtime before her name came up again. At 8:00 I received a phone call from the nurses on the orthopaedic floor. Melissa's heart rate had spiked up to 140 about eight hours ago.

Eight hours ago? Why wasn't I called? Her medical team ordered a cardiology consult, and the cardiologist diagnosed *tachycardia*. Generally anything above one hundred beats per minute is tachycardia, a heart rate exceeding the resting rate. A fast heart rate was to be expected after surgery. Her body had a lot going on. But 140 is very high.

"Any other issues?" I was scrubbed in surgery, so I had to relay messages back and forth through the nurses.

"Well, the only other problem was that she ran a fever of 103 at 5:00 a.m."

To some extent, this could be expected. Surgeons realize it's ordinary for those recovering from surgery to run a low-grade fever, that is, not exceeding about 100.4 degrees. It's called *atelectasis*, and it's a result of gases remaining in the lungs from anesthesia. Melissa was at 103, and that's too high. If you've been put to sleep for surgery, you might recall being asked to cough and take deep breaths after you woke up. That's so you can push out as many of those gases as possible. The team had taken blood cultures and given her acetaminophen, which had helped—her fever had dropped to 99. I said to keep a close eye on her during the morning, and that was that—or was it?

Five minutes after the call, I couldn't get Melissa's condition out of my thoughts. Something was troubling me—something in the back of my mind. This wasn't anything coming out of medical training or logic; it was pure intuition, if not something more mysterious than that. I had a strong sense that Melissa was going to crash and become septic—her full body becoming infected through the bloodstream.

Septic is the word we now use for what we once called "blood poisoning." Bacteria goes out to the full body, using the bloodstream as train tracks, stopping at all the main stations—liver, kidneys, heart, and other organs. At every stop bacteria cells unpack and get busy. Security agents in the organs should keep this from happening. But if the systems' defenses are down, and the security agents are not strong enough, the organs begin to collapse. Sepsis can happen quickly and become fatal.

It wasn't enough just to accept that Melissa's fever had dropped lower, closer to normal. The problem was, we didn't know why it was so elevated in the first place. And as I've been taught, *not* knowing is the worst position for a doctor. There was too much not knowing in this case. High fever, rapid heart rate, unidentified bacteria—I was uncomfortable with a growing pattern of conditions that lacked explanations.

I gave orders to move Melissa to intensive care (ICU), and I would take her into surgery in two hours. I gave specific directions on what to do and why I was concerned about Melissa going septic.

I went to the ICU as quickly as I could and saw Melissa's husband and her father in the family waiting room. After I'd explained to them what was happening with her fever and heart rate, they wanted to know why since this past surgery had been so much more promising.

"We can't be sure. We've given her five of the strongest antibiotics, and her arm looked really good yesterday. But these new developments suggest the infection is spreading to the rest of her body. I'm about to check those latest cultures and see if this time we're able to identify any of the bacteria."

I sat down at a computer terminal nearby, logged in, and found the results: *no organisms growing.* I slumped into my chair, more puzzled than ever. How was it we had such an obvious and serious bacterial infection and *none* of it was leaving a trace?

It also meant we'd need to add yet another agent to cover the possibility of a fungal infection. For the next twenty-four hours, our goal would be, first, to keep her organs from shutting down and, second, to get Melissa back to a normal temperature.

In her room Melissa's monitors showed a heart rate in the 120s and a fever of 103. But when I looked at her, she calmly asked me, "What am I doing in the ICU, Doctor?" She didn't seem to feel very sick.

I explained to her about the possibility of becoming septic. "Oh, well," she said. "I'll just hang out here, then, and wait for my body to bring the fever down."

Jeff Miller was the ICU doctor for this case, and he was a good one. We've worked together for many years, and he agreed with my approach. So an hour later Melissa was back in the operating room—her seventh visit. She was put to sleep; then I removed the few sutures I had placed and began washing the arm with sterile saline. I couldn't see any difference from the day before, when her arm showed no sign of infection, though there were small areas where pus was beginning to collect again.

Once again I took cultures, washed out the visible areas of pus, and removed a small amount of dead muscle. And, again, I left the wound open. We wrapped her in more sterile dressing, woke her up, and returned her to ICU. Melissa wasn't sure why she was bouncing around from a regular room to ICU to the OR and back to ICU, but I explained it as best I could.

This woman was running a high fever for no discernible reason. Plus, the diabetic angle complicated everything. Normal blood sugar levels run between 80 and 120. I was worried about her blood sugar rising from the infection.

I hoped I could trust myself. I was experienced enough, but this was the kind of case no one can prepare you for, and I felt the full burden of needing to make the right decisions and help Melissa keep her right arm.

16

INVISIBLE ENEMIES

That afternoon, while spending time with other patients back at my office, I felt a sudden urge to check on Melissa one more time. It wasn't absolutely necessary. Presumably somebody would call me if there was an issue. But *presumably* wasn't enough; nothing we'd presumed about this case had worked out so far. So I intended to listen to my instincts. I returned to the hospital and proceeded directly to the ICU, where I greeted Neil and Melissa.

Nothing had changed. Fever: 101 to 103. Heart rate: 120 to 140. Both were too high for too long. It occurred to me I hadn't talked much to Melissa before that day's surgery because I'd been hurrying, trying to get her stabilized. Now I grabbed a chair and talked to the couple.

They wanted to know what was going on with these symptoms, and I had to use those three least-favorite words of every doctor: "I don't know." This family didn't need me dispensing false hope. I brought up my blood sugar concern.

"Melissa," I said, "I'm worried about the diabetic angle of these things. By any chance, have you been keeping track through your insulin pump?" It gives readings that make that possible.

"Absolutely," she said. "I write them down three times a day. I can show you my log." I was grateful for this information—I could have gotten it through the records on the computer, but it would have required lengthy navigation through a world of various screens.

Melissa reached over to her dinner tray and picked up a napkin. I said, "Listen, go ahead and finish your dinner. You can show me your notebook and documentation afterward."

She said, "Oh, no, I'm not using a notebook. I kept track of my blood sugars on this napkin."

I couldn't believe that. "Melissa, news flash: if you ask the nurses, they just might give you an actual pen and a pad of paper. Are you telling me you've been keeping track of these important figures on a brown napkin that someone might crumple up and throw away while cleaning up after you?"

"I guess so," she said and grinned. "But it hasn't happened yet." She handed me the napkin. I glanced over the figures and found the sugars were surprisingly well controlled. The highest was 150.

"This is strange—in a *good* way for a change," I said. "Given the infection you've been facing, these numbers are really fairly reasonable." I handed the napkin back to her and said, "May I ask you to copy these to an actual piece of paper so we don't have to go looking for them in the dumpster out back, in the middle of all the meal remains, sometime later?"

She smiled. "Sure, I can do that."

As I walked out of the room, I felt a wave of optimism, halted, and spun around. "Melissa, you're going to be all right."

"I know. I'm not worried—I have you taking care of me." And she offered a calm and confident smile.

I should have considered that in tandem with the more mysterious aspects of her case—then I might have made the connection sooner. I might have thought, *There's more than chemistry and biology and all these natural processes going on here. This is an uncommon case* and *an uncommon patient. There just could be something going on here that is beyond our comprehension.*

But you see, my mentors didn't train me to think that way. Nobody in the medical field is taught to look beyond the carefully tended walls of science—they'd be laughed out of the academy.

But a large number of doctors and nurses know better. We've seen things that charts and tests can't measure. We've seen those occasions when prayer is as good an explanation as anything else because, frankly, there's no reasonable explanation that works. Sometimes it's simply a patient's powerful will to be healed.

Still—even knowing that—our training and the daily, mundane course of life lull us into believing life is more predictable and quantifiable than it really is. How often have you heard the phrase "You can't make this stuff up"? It's because truth is often stranger than fiction or, at least, stranger than our expectations. Certain situations and people defy the normal logic.

This was one of those situations, and Melissa was one of those people.

With all the mystery and uncertainty, that strange, gut-level intuition was telling me to be encouraged. I'd look at the numbers and feel despair. Then I'd talk to Melissa, and my spirits would be lifted.

After this we saw a slight elevation in a marker related to her kidneys. The organs were overworking. As tiresome as it had to be for Melissa, this meant one more trip to the operating room for the same surgery I'd been doing—debriding, which refers to the removal of muscles that have lost blood flow.

This time, however, I saw what I desperately did not want to see: she was slowly losing viability in her forearm muscles after all. I thought back to the second Sunday procedure when I'd seen the pinkness in those two small areas—at the time, it wasn't so bad. But her defenses were wearing down. Her body could take only so much. It was fighting hard, but it was wearing out while the enemy just got stronger.

During this point in the surgery, I was told her blood count was slowly dropping—a new problem. With all the challenges, we hadn't seen anything going on with her red blood cell count. I was getting closer to despondency when it came to Melissa's future. We had talked in terms of saving an arm, but I recognized—without wanting to face it—the growing likelihood that the stakes were larger than that. *Much* larger. Melissa's life could be in jeopardy.

We gave her two pints of blood. I continued debriding the dead muscle. She was definitely losing the main muscle group used by the body to move the fingers.

I looked at the eyes of my assistants above their surgical masks, peering back at me as I worked on Melissa's muscle. I saw the flat gaze of defeat. Everyone who takes part in an operation hates that moment when an inescapable, heartbreaking conclusion becomes evident.

"I know what you're thinking," I announced to the room. "But I'm not there with you yet. We can still save this arm."

They wouldn't meet my glance. Later, looking through the intraoperative photos, I realized my will and my emotions were speaking rather than my scientific side. There's no way I could have held the *objective* opinion this arm could be saved. But that's as it should be. You want your surgeon to err on the side of fighting for you and your future.

For the fifth time I took cultures, washed out the arm, and returned her to the ICU. She was getting to be as much of a veteran of this procedure as we were. Everybody knew her by now. But her situation didn't improve. It appeared that the infection was making steady progress in spreading throughout her body.

We gave her two more pints of blood. Where was it all going? It had to come from her arm, but she'd lost little of it in surgery. That raised the grim possibility of bleeding in internal organs. Thankfully, a CT scan showed no problems there, no blood in her urine, no bleeding in the brain.

I talked to other physicians and heard various ideas about the disappearing blood. The most likely one was that her red blood cells were being destroyed in the bloodstream. If that was true, we should have seen traces of it in the urine. We didn't. Another invisible attack, another unexplainable symptom. It was as if her body was working on principles other than scientific ones—its own set of rules.

Melissa received twelve pints of blood over forty-eight hours, and we still couldn't keep her hemoglobin level above 7 (normal would be 14). If it dropped as low as 3, I knew her life would be in jeopardy. There would no longer be enough red blood cells to transport oxygen to the organs.

Fighting invisible enemies was exhausting for me; I couldn't imagine how it felt for Melissa. She was the one whose life was at stake. And, yet again, no bacteria was identified in the cultures. Incredible. Melissa had not only me but a small army of other physicians and experts puzzling over why she was becoming septic. I was seeking any and every suggestion to help her become stabilized.

I was always careful to keep Melissa's family in the loop. They understood just how much danger she was in. Full sepsis meant long-term complications *at best.* I tried to be encouraging. "I'm just not going to let a patient come away with complications after a raccoon bite to the arm," I said.

"Please do what you can for Melissa," they all said, in their various ways. Michelle, Melissa's sister, was especially broken up and given to tears. I hugged her and told her we were still fighting, and we didn't plan to quit.

17

PREPARING FOR THE WORST

I ran into Dave, Melissa's brother-in-law, in the hallway. He was often the first one to come forward with a question. He took me aside and said, "Doc, I want to thank you for everything."

"Well, I wish I could be bringing you much better results."

"If you can't, I'm sure no one else could, Dr. Seth. Since this whole nightmare began, you've done more for Melissa than anyone else. You've been there for her. As a matter of fact, will you thank your wife on our behalf? Because we're not sure when you go home. You're here in the morning and you're here at night. We see how much you care and the courage you give Melissa."

"Her courage is all her own, Dave. I don't get any credit for that. I've never had a patient quite as strong as this one."

"Well, you've encouraged us all along, but we appreciate the way you've done it without ever sugarcoating it."

Those words, together with Melissa's ironclad trust, only served to humble me and make me more desperate to do what I could for this family. I knew I was no more than an ordinary doctor, and I had no medical superpowers to offer. But I had bonded with this family because of the special conditions of this case.

I think they knew I hadn't been sugarcoating her condition or throwing out empty words to soothe their worries. I believed everything I told them. I was speaking based on my long experience with this kind of case. I would never in my wildest dreams have expected us to end up in the ICU with the patient's life on the line.

Despite what Dave told me about my use of time, I felt that I hadn't talked much with Melissa. For one thing, she was often medicated or sleeping when I visited. Also, I had so little good news to bring her. And, to some extent, I was trying to guard that courage that might have been her strongest—even her only—defense. She needed to know what was going on, but I didn't want her knowing the worst at every possible moment and overthinking it. The *a*-word—amputation—was one I really wanted to avoid in our conversations.

But although a lot of this period is now a blur in her memory, the days fading together and meds obscuring a good bit of her recollections, she really knew the score. Melissa was smart enough to know that when her heart was racing and her fever remained high, she was in a dangerous situation. Then, when her kidneys started to have problems, she realized that it was her whole body, not just her arm, under attack. She understood that a powerful infection could be deadly. And she knew that, in the long run, it was better to hang on to a life than an arm.

Yet she didn't complain about any of that. While I worried about broken promises and an animal bite I'd told them was just routine, she was the one encouraging me. "You're doing everything you can," she said. "If there were an easy cure, you'd have already found it." When I told her about a new course of action, she accepted it immediately, with no sign of doubt in her eyes.

Within her private world she wasn't quite so sunny. I learned later that when I told her that her kidneys and liver were shutting down, she had begun to entertain thoughts of death. Few of us are seriously asked to do that in our early forties. She squarely faced those two crucial days in late August as her last on earth. Neither Neil nor I knew this at the time because she didn't share it.

This is what death feels like, she thought. Melissa doesn't wear her deepest thoughts and feelings on her sleeve. For two days she walked through the valley of the shadow of death alone, by choice, because she doesn't like to trouble other people. So the courage and the calm, the unruffled spirit I saw, weren't the whole story. She certainly knew her life was in peril, and she wasn't ready to die. But she faced mortality on her own terms.

She told me she knew the time was coming when an amputation was the only choice. Others understood it too. "Let's do it and get that infection cleared," people said, almost as if this was one more item on a to-do list.

But up to the very last microsecond, I searched in vain for some other way. I thought about Melissa's life with one arm. She wouldn't be able to write the way most of us do, with one hand to hold down the page. She wouldn't be able to tie her shoes. Embracing with one arm isn't quite the same—neither is playing with a dog, for that matter. Imagine trying to cook or get dressed or work in the garden.

There are thousands upon thousands of little actions in life we take for granted, as long as we have both our hands for those things. There were prostheses, sure, but I didn't know too much about them at this point, other than that none of them offered anything close to having a human, flesh-and-bone hand.

Losing a hand is a price worth paying to save a life. But that doesn't mean it's not a cruelly high price. I focused on the current antibiotics as if they were magic potions, and I willed them to come on strongly, drive out the infection, and allow her body to begin to heal in all the places it was ailing. At this point it would be a miracle, and doctors appreciate miracles just like everyone else.

Melissa was more realistic about these things than I was. As I found out, she was already over it—the loss of her arm. She'd been focused, up to a point, on saving it, but now she was more interested in staying alive.

She was thinking, *I'm not ready to leave my husband. I want to see my dogs, and I want to keep helping stray dogs find homes. I have things I still want to do even if I have to do them one-handed.*

I'd made a promise to this family—I was their champion, the knight who would go in to fight the dragon. And the dragon was whipping me day after day. I don't like to be beaten, not even when playing basketball with my son, Jaideep, and I couldn't accept defeat when it came to this case.

I checked her labs. Not only were her kidneys starting to shut down, but now her liver wasn't looking good. The clock was ticking. I had twenty-four hours, and unless something changed, I would be breaking a huge promise.

I would be amputating an arm.

18

"NOW SAVE MY LIFE"

The next day, August 26, 2015, will always stand out as a tall monument in my memory. As I awoke that morning, I felt deep inside me that this was going to be a day when decisions were made that determined futures.

I found Melissa was stable after the night—no worse, no better. Wasn't that good news of a kind? It could mean the new antibiotics were working, stopping the sepsis in its tracks. Maybe she was turning the corner. I crossed my fingers, expressed optimism, and told Melissa I'd check back in with her after my morning surgeries. Some of my colleagues would be evaluating her condition in the meantime.

Yet this day continued to carry the feel of impending doom. During spare moments in my operation, I checked the clock anxiously and hoped I wouldn't be advised about an emergency call—that would mean things had turned for the worse with Melissa, and her body was at the early stages of shutting down.

I'd been operating for five hours now, and no news was good news. Then I heard the surgical nurse answer the phone. "He's operating right now," she said to the caller.

"It's not a problem," I told her. "Who's on the phone?"

"Dr. Miller from the ICU."

The nurse held the phone to my ear. I said, "Hey, Jeff. What's going on?"

"Ajay," he said firmly, "if you don't amputate that arm within the next hour, she'll die."

My heart rapidly and utterly sank. Deep down I hadn't allowed myself to believe we were really at that point.

"I'm on my way," I said flatly. I had the surgical nurse call and get an OR for Melissa for an emergent amputation. The hospital was very busy, and many of the rooms were booked, but I sent instructions that this was life or death—the arm had to come off, from above the elbow, within an hour.

The scheduler reserved the room.

I was, as usual, in a different hospital. I hurried out to the parking lot, pushed my car through the traffic as fast as I could without endangering lives, and kept finding myself thinking back to that first occasion when I met the woman with the raccoon bite. I thought about my supreme confidence—I'd almost call it swagger—as I assured them I'd never had to amputate a limb after this kind of bite. It was a small animal bite. What could go wrong?

I'd asked for, and received, the family's trust. After that, Melissa had been through surgery after surgery, and somehow it seemed as if her condition was worse after each one. It wasn't as if we were giving her poor care or causing the decline. I just wanted to see my patient healed. I wanted to bring good news to this family for once.

I bypassed the nurses' station at the ICU and hurried directly to Melissa's room. There I met eight people from surgery, whom I'd called on the way. All of them wore grim expressions that matched mine. I greeted them and entered Melissa's room. The nurse was readying her for surgery.

I smiled at Melissa, gripped her hand, and tried to be encouraging, but I found I couldn't speak. I had to turn my back and consult the computer. There was nothing on that screen I really needed to see; I didn't want to show her my face. My eyes were wet, and even with the surgical team there, my mind was scrambling for some alternate route, some way to save her arm and the daily life she knew.

I swallowed my emotions and looked closer at Melissa. She was, once again, the picture of calm. What should I say? How could I talk with her about losing an arm for the rest of her life? "Melissa," I said gently, "I'm sorry. I'm so sorry I couldn't save your arm." Every one of those words was like a hot coal on my tongue.

I met her eyes and saw trust, somehow—the full trust a patient places in a doctor. IV lines were attached to both arms. Compression stockings were on her legs, and an array of monitors beeped as various personnel hurried back and forth.

"You did your best to save my arm," she said. "Now, please save my life."

"Melissa," I said. "You're not going to die. I won't let it happen. I promise."

We discussed the process in short bits because she was in and out of consciousness at this point. Once she was fully sedated, I turned to her husband and said, "I've got this, Neil. I *will not* let her die."

He nodded, but I could see the fear in his eyes. Who would have felt differently? And why should he believe me at this point? I had the family's trust, but they understood Melissa's condition was deadly serious.

The team began wheeling Melissa Loomis toward OR12. I lingered for a moment in ICU and wondered if some other doctor could have done a better job. But I knew I'd done not only everything in my power but everything any other doctor would do. I felt I was in the position God wanted me to be in, and that I needed to be courageous and stay the course for Melissa's sake.

As I walked by the nursing station, I saw Rhonda in tears. She'd been a nurse in three of Melissa's surgeries. "I'm supposed to be the circulating nurse," she told me. "Please understand—I'm not emotionally capable of doing this one."

Her reaction was typical of everyone in the hospital who had gotten to know Melissa and been a part of her care. Rhonda needed to be cool and objective, but by now it was like operating on a very close friend. Saving the arm had been something of a crusade for those who knew and loved this patient. The case was taking its toll on all of us.

I understood. But the surgeon is the one person on the team who can't opt out. I was emotionally involved too; the last thing I wanted to do was give in to an amputation. But all other roads were closed. This was the only path that lay before us.

I quietly scrubbed; put on gown, gloves, and magnification glasses; and entered OR12, prepared to remove a good friend's arm.

19
AMPUTEE

What is there to say about this terribly sad form of surgery?

In our country the largest number of amputations occur because of complications of blood flow—especially in the case of diabetes. Melissa was a diabetic, but it didn't follow that this was the root of the problem for her.

All we knew was that her system was entering the most dangerous stages of shutting down, and removing the forearm was the one strategy, the one possible fail-safe way to eliminate the problem.

After we put her to sleep and prepped her arm, I could see how much muscle in the infected area had died off, not only from the infection but also due to a lack of oxygen. The plan was to amputate above the elbow, transecting the nerves to the hand.

I made my incision, removing the now-useless muscle and bone. It wasn't an incredibly complex or time-consuming operation—just a heartbreaking one.

After finishing the operation, I made my way to the waiting room. As I spoke to the family, I couldn't help but hear in my head the promise I'd made to them. There was no way she'd lose that arm, I'd said. I needed to learn there's *always* a way. Nothing involving life is certain.

I didn't grin or crack jokes or offer a big thumbs-up when I met the family. I did say, "It went well. We did what we had to, and over the next twenty-four hours we're watching for her bodily systems to rally and begin their recovery. Let's all keep her in our thoughts and prayers, and of course I'll let you know everything we find out, as quickly as possible."

I embraced each one of them and left for my office, where I'd finish up dictations and paperwork. I had an enormous sense of finally concluding the major portion of Melissa's story and of how, in eighteen hours, I would start the day with another group of patients. I needed to keep my spirits up, focus the way I always had, and be sure each new patient received the best I could offer. A doctor must keep moving on.

I had to devise a new game plan. I would also check up on Melissa at 6:30 a.m. If I got through the night without any news about her, that was a good sign. It would indicate that the source of the attacks on her body was gone and her systems were regrouping.

And what if that call did come? What if the amputation still hadn't solved the problem? Nothing had gone as expected so far, so why should I be confident? We still had no idea what kind of infection this was. Maybe it would live on until Melissa could not do so. Maybe it would demand her very life.

I couldn't afford to think like that; if I did, the infection would have worked its way to my mind, my soul. We can't allow our spirits to be poisoned.

All the same, the prospect of checking on Melissa filled me with hope and dread all tangled together. I had a particularly quiet day in my office after performing her surgery, talking to fewer people than I usually do before leaving for home at the end of the day.

Keeping my work at the office as I do, I understood my kids had no clue what I was feeling, and that was for the best. It was just another day at the office for Dad. Over dinner I asked them how things had gone for them, and then my son and I played basketball. It was a needed outlet for all my nervous energy.

"That's it for me," I said after a while. "Let's go wash up."

"I crushed you," Jaideep said, taking one more shot at the hoop.

"Oh, is that so? I keep score by number of shots that touched the rim, and by that count, I won."

"No way!" he protested. We both were laughing as we walked back into the house. It felt good simply to goof around, if only for a few minutes.

But that night, needless to say, I didn't sleep well. I stared at the ceiling and wondered how high Melissa's fever was, whether her kidneys and liver were stabilizing, and how she was processing the emotions of losing an arm.

I rolled over and stared down at my cell phone beside the bed. I could call the ICU and get the numbers.

In the end, I didn't. Maybe I really didn't want to take the chance of hearing bad news. If that happened, after that surgery, I'd have no more cards to play. I told myself the ICU staff had things under control, and if there was anything I should hear, they'd call me.

Sometime after that, I drifted off. Then I cursed the alarm when it began to blare obnoxiously at 5:30 a.m., what seemed like ten minutes later. I wished for those missed hours of sleep.

I was tense as I arrived at the second floor and passed through the ICU doors. I took a deep breath, pushed aside the curtain, and prepared myself for the sight of monitors, IVs, and charts that still had nothing positive to say.

Instead, I saw the last sight I expected: a fresh amputee with an expansive, good-morning smile. Melissa said, "Hey, how are you? Seems like the surgery went well. You want some eggs?"

I felt a significant portion of the burden slip off my shoulders, and I could breathe again. I even smiled back.

Where do patients like this come from?

20

ANOTHER DAUGHTER

Have you ever noticed how long journeys make for lifelong friendships? Melissa, her family, and I had just traveled such a journey. We shared some of the worst moments, as well as a few nice ones, and we bonded during that time. In the midst of the struggle, I found myself opening up with Melissa and sharing a part of my life I'd ordinarily never discuss with a patient.

She was experiencing the heartbreaking loss of a forearm. Somehow I found myself telling her about my own greatest loss. Perhaps I wanted her to understand that I knew what it was like to grieve. Or maybe I opened up simply because she'd become a close friend.

"I've told you about having two children," I said to her. "Jaideep and Trinity. But that's not really the truth. Actually, I have three children. There are two who live in my home, and another who lives in my heart."

Melissa raised an eyebrow, making the connection, and I began to tell her about Jayani.

Jayani Asha Seth was born on Christmas Day 2004 at 6:30 a.m.—the most wonderful Christmas gift I ever received. She was a beautiful and completely healthy child, and she brought our family all the joy a new baby has to offer. Topping off my first year of medical practice, the new addition put me over the moon. I was part of a new practice and had a beautiful new daughter at home. It was a wonderful time for me to be alive.

Five months later the four of us took a trip to Disney World. Our son, Jaideep, was two, and Jayani was an infant. Travel is more difficult with a newborn, but we felt our baby was ready for her first airplane trip. Jaideep was all about meeting Mickey Mouse. That was number one on his to-do list.

We had a terrific first day at Disney World, then waited for the bus to take us back to the hotel. Kim and I talked as she held Jayani, and Jaideep ran circles around us, as two-year-olds will do. "I know someone else who's coming to Disney World soon," I said.

I'd told her about a patient of mine, a little boy diagnosed with leukemia, whose grandmother had contacted Make-A-Wish. And sure enough, in a month this little boy would live out his own dream of visiting Disney World.

"This got me thinking," I said. "We've been blessed with so much during this year. We're both fortunate enough to have rewarding careers with medical groups that help people, and we have two beautiful, healthy children. But we could help in other ways, particularly financially."

"Yes! I've been thinking about that too."

"I'd like to do the same thing Make-A-Wish did for that little boy, only with another child who's facing a serious cancer diagnosis. When I see a little boy or a little girl who is dying of cancer, it breaks my heart because nothing can be done to save that life. One thing we *can* do is to make one week of that life perfect. I can't imagine a better investment of our money."

"That's a great idea," said Kim. I'd known it would be right down her alley. "Let's call Make-A-Wish when we get back to Ohio, and we'll set it up."

We were still feeling good about that conversation the next morning when we noticed that Jayani was showing signs of illness. While babies do catch colds and viruses, especially in crowds, something seemed a little frightening about her symptoms. She was showing signs of quickly going downhill. A sick five-month-old is a deal-breaker for vacations. For some reason, we both knew this was worse than the normal stomach flu. It was only the second day of our trip, but we caught a plane and left early.

Back home, we took Jayani to our pediatrician, who checked her out and admitted her to the hospital. Our child was diagnosed with meningitis, which meant her brain and spinal cord membranes were inflamed and she would be on several days of antibiotics. But twenty-four hours in, when we should have begun to see improvement, things were no better. Kim, being a physician herself, asked for a CT scan.

Kim was in the room with Jayani during the scan while I stayed with the radiology technician who was operating the X-ray equipment. What happened next made for a terrible moment—medical life and personal life colliding.

As the scan progressed slowly from the top to the bottom of the brain, I noticed on the screen a large mass. I told the tech to return to the previous image. The question still rings in my ear: "What is that . . . what *is* that?"

I knew at that moment; she had a brain tumor.

Jayani was taken to the ICU at Akron's Children's Hospital, and we discovered she had a malignant ependymoma, a brain tumor rising from the central nervous system and extremely rare in such a small child.

A hard year of surgeries and chemotherapy ensued. Halfway through, we traveled with Jayani to St. Jude Children's Research Hospital in Memphis, Tennessee. If you happen to be a financial supporter of St. Jude, thank you. And please know you're making one of the most helpful donations imaginable.

The hospital gave us wonderful care. But on May 30, 2006, Jayani Seth lost her battle with the tumor—the worst day of my life. Losing a child isn't a level of suffering that can be put into words, so I won't try. You can only hope and pray you never come to understand the depths of that kind of misery.

While Kim and I grieved, we received wise words from my colleague Dr. John Riester: "I can't begin to imagine what you're feeling right now. The one thing you must do, however, is make sure you don't ruin your two-year-old's life. It wouldn't be fair to him."

The truth of that really hit me in the face, and the effort to live up to those words got us through those next few months. We lost a child, but we still had a child, and that one needed us as much as ever.

Somehow, life moves on, cruel and unfeeling as that seems. The world spins the course of another day, then makes one more revolution around the sun. It has made twelve journeys around the sun since Jayani left this world, but not a day goes by that I don't think of her. That's why I say we have two kids in our home and one in our hearts.

I told Melissa this story just as I tell it to you. What gives the most meaning to my sense of loss is that my daughter lives on through the inspiration she gives me to help patients like Melissa, and also through Jayani's Army (www.jayanisarmy.org), our nonprofit organization that helps children with cancer.

We discovered that coping with your child's cancer is a lonely ordeal. We all fight better when we're part of an army. Jayani's Army enlists us all to combat the enemy of cancer, and the advances we've made on that enemy, as well as the comfort and support we've been able to bring, have been our way of turning mourning into joy.

Melissa knew that Jayani was close to my heart during the key moments of our journey, and the greatest example of that was yet to come.

PART 3
THE PROMISE

21
A BIRTHDAY CAKE

An amputation, sadly enough, is a forever thing.

Which is why we make absolutely sure there's no other option before removing a part of the body. Arm and hand transplant procedures are rare and extremely problematic. Melissa's case was more typical. She knew it was time to learn about life with only one arm—assuming she could get her general condition back to normal. She had a long way to go.

I watched her like a hawk over the next few days. As she began to turn the corner, just after her surgery, I was deeply encouraged. Melissa's improvements came quickly and surely. Liver function: stabilized. Kidneys: showing progress. Heart rate: back to normal. Melissa was finally showing genuine improvement.

By August 29, we were able to talk about letting her leave the ICU and return to a regular room. She was delighted because she knew it meant she was one step closer to going home. She'd been in the hospital for more than two weeks—far too long—and it hardly seemed possible to her that she could sleep in her own bed and see her dogs again. We take those things for granted until they're out of reach.

The only item missing was an identification of the bug causing her infection. It still remained a mystery. We thought the five antibiotics would be enough to drive it away once and for all. Continuing to receive those drugs was all she had to occupy her time in her new room—that and chatting with her visitors, as well as making friends with nurses, something easy for her to do.

So when I visited her early on the fourth day after her amputation surgery, she wanted to know exactly when she would be able to get out of the hospital. I'm sure it was beginning to feel more like a hotel for her.

"How does your birthday sound?" I grinned. She'd told me that on the first of September, she'd be a birthday girl. It was cautious timing from our perspective since I knew her condition could change in the blink of an eye. But hospital release seemed like a pretty good birthday present for a patient who's tired of the hospital.

The big day arrived. All signs were go. She was packing her things, excited about leaving, when I entered the room. She was chatting about all the things she looked forward to doing at home. That's when we took her temperature, and it had spiked upward—right on cue, as if life had simply decided she couldn't escape the hospital.

We each took a deep breath and tried to absorb the disappointment of the moment.

I was utterly flabbergasted. Emphasizing the day as her birthday just made circumstances crueler. More than any patient I'd known for a long time, Melissa deserved something very good to happen to her. I remember thinking that we were so close to letting her see the outside world again. We were going on Day 16 of her hospital stay, and I decided we'd just have to make the day special by any means necessary.

I couldn't send her home for a celebration—but who said I couldn't bring the celebration to her? As soon as I had a spare moment, I drove to the nearest grocery store, walked over to the bakery section, and said, "Hi, how are you? I'm a surgeon over at the hospital."

The girl smiled and said, "Hi! How can we help you, sir?"

"Well, I need a birthday cake for one of my patients. But I'm not talking about your typical birthday cake. I need an awesome cake, one that comes with smiles guaranteed."

"Our cakes do that! Anything we bake would make *me* smile. But let's see if we can come up with the 'awesome' part. Anything in particular you want written on it?"

I had a spontaneous inspiration, and I knew it was either ingenious or one of the dumbest ideas of all time. "I think I'd like it to say 'Happy Birthday, Melissa,' in Chinese characters."

The girl registered a blank expression. Then she looked over at the baker, who had just come out from the back. He joined us, his hands on his hips. I was thinking, *Okay, this idea was from the "dumbest ever" variety.*

"Oh—wait!" I improvised. "I should have told you this patient of mine is a young lady who came from China. She's contracted a rare disease, and our hospital here in North Canton is the global authority in curing it."

Okay, now, that's not absolutely true, whispered my conscience.

As a matter of fact, I told my conscience, *it's not at all true. It's a fairly sizable whopper. But look at them—they're actually buying it!*

Hey, you're right, said my conscience. *Go for it, man.*

I generally let my conscience be my guide, but I've already told you: I was going to bring a smile to Melissa's face by *any means possible.*

The rare disease idea caused the girl's eyes to light up. She looked at the baker, then back at me, and just like that, she was on a mission. This was a VIP cake! A Very Important Patient, with a rare Chinese disease that could be cured only in this very corner of the Buckeye State.

Except there was a problem. "Umm. Nobody here knows Chinese," she said.

Shocker, I thought. I pretty much knew that, but I was already on it. One thing I've learned, above and beyond the medical world, is that you can google anything. "Happy Birthday, Melissa, in Chinese," I tapped on my phone.

Google can't be stopped. I handed over my phone, which read,

生日快乐梅丽莎

She studied it; the baker looked over her shoulder, squinting. And just like that, she was looking for a cake, and he was hurrying to grab his icing pencil. It seemed that making a Chinese birthday cake was the coolest job they'd *ever* had. I specified that it was absolutely mandatory that the cake be chocolate—no vanilla. I think they felt this had something to do with the rare illness, but the truth was, I hate vanilla.

The bakers were creating a confectionary masterpiece, something they'd tell their grandchildren about. I had to walk over to the produce section so I wouldn't burst out laughing in front of them. (Now I guess I have to find a way to keep them from reading this book!)

Five minutes later they had a beautiful cake with Chinese characters on it. I bowed to them like a good Chinese surgeon and was soon driving the cake back to the hospital, along with the bundle of balloons I'd bought. One of them said "Happy Bar Mitzvah," and another, "Congratulations on Your New Baby Girl!"

I parked the whole spectacle just outside her room, then entered with a somber expression on my face. Melissa and her family said hello.

"I know what you're going to say," said Melissa. "I have to stay here on my birthday, don't I?"

"I'm afraid so. Your fever came back just before we could let you go. But there's one more treatment we can attempt."

"Oh? What's that?"

I had somehow kept a grim countenance. I turned, went to the door, and grabbed the cake and the balloons. Then I bustled back in, holding those crazy balloons with the inappropriate messages and a cake that spoke Chinese. She and her husband began laughing.

"Well, I'm glad you brought me this stuff," she said. "But what I'm *really* glad about is that I know you well enough not to think you're crazy."

"Glad to hear that. I'm not so sure, myself."

The family chuckled over the cake as the nurse left to get plates and forks. You could feel the mood lightening.

"You guys go ahead," I said. "I need to check on some lab results." I went to the nearest computer and tried to figure out why Melissa's fever was coming back. Worse, it looked as if her blood count was falling again. Always more questions. Whatever we were fighting, it was serving notice it wasn't going down without a fight.

"Melissa," I said as I reentered the room, "you're going to need some more blood. I also think we'll need to rewash your upper arm at least one more time, in surgery, and change your antibiotics." She nodded. She'd been through it so many times, it was almost a what-else-is-new? moment.

"One other thing," I said. "Will somebody here cut me a piece of that cake? It just so happens I *love* chocolate."

"I'm so sorry! We ate the whole thing." Melissa placed a hand over her mouth, covering an embarrassed smile.

My eyebrows shot up, and I put on my dead-serious face again. "Let me get this straight. I went all the way to China to get you a cake, and you didn't even save me a piece? Is that how you reward the surgeon who saved your life?"

"Just hang on till tomorrow," she said. "We'll bring you some cupcakes."

Neil said, "I'm on it."

I mulled over that offer for a second. "Cupcakes? Hmm. They'll have to do. But if I come in here with a big appetite tomorrow, and there are no cupcakes, or if they're not chocolate, we're going to have some serious issues."

"We always listen to our doctor."

We were all in good humor as I said goodbye, but deep inside I was upset about everything she was having to go through. Laughter may be good medicine, but I preferred the kind that solved a physical problem.

22
WONDERFUL, DANGEROUS HOPE

Melissa's fever continued to hover between 101 and 103 degrees over the next few days. And she was steadily losing blood for unknown reasons. There was really no option apart from another trip to the OR for rewashing the amputated limb and giving her another two pints of blood. Melissa offered no objections. "I trust you to do the right thing," she said. She had the faith of a child, the strength of a Roman warrior, and the courage of a military hero.

During surgery, I really didn't discover anything that would account for her fever. A small amount of muscle wasn't getting enough blood supply, but no bacteria seemed to be present that needed to be debrided. As usual, there were no ready answers.

I lay awake that night—which was getting to be a habit—and reviewed every aspect of her care. I racked my brain to come up with an explanation of why she was losing blood and continuing to have a fever. I'd searched every part of her body that could possibly account for blood loss or explain the fever. When I have an unsolved problem, I have trouble moving on. That's how it is for doctors who are committed to healing every patient. We just can't let an illness beat us.

The other issue was that until we handled this problem, we couldn't move on. Melissa couldn't begin to get her life back. She was dealing with a missing arm without the positive therapy of exploring new options. I wanted her to be able to focus on the future in an encouraging way. For now, we couldn't even close up the stump—not until we found out what was wrong inside.

Soon after her birthday I entered the room to find Melissa sitting on the end of her bed. Neil was hunched over, looking at something on her phone.

"How's everybody today?" I asked.

"This is what I want," said Melissa, eager to show me something. She thrust the phone into my hand, and I saw she was watching a video posted on Facebook. As I pressed Play, she said, "Right there is the type of arm I'd love to have."

The video showed a three-dimensional model of a prosthetic arm that could move every joint, fully opening and closing—the next best thing to a real hand. "Think of the things I could do with a hand like that," she said.

This was good—very good. Melissa was thinking about what she could do rather than what she had lost. That's a strong recipe for recovery.

Neil said, "Now that her arm is amputated, what are we looking for to replace it? What kinds of prosthetic options are there? This is just an animation. Is something like this prosthetic arm actually available?"

They were, of course, completely new to this subject. No one ever plans to lose a limb. They played the video over and over, their eyes focused on the screen as they studied how lifelike a prosthetic hand could be. Melissa was wearing the biggest smile. She was illuminated from within by the hope of going home, reentering her daily life, and learning to use a futuristic arm.

As her doctor, I was delighted to see her attitude; but I also had a fever to figure out. We had to be absolutely certain the mysterious infection was eradicated, and then we had to get her moved back home. After all that, we could start thinking about that prosthetic.

I hadn't focused at all on this subject for that reason. Yet I watched Melissa's fascination with the video on her phone and how it was energizing her. Something made me say, "Melissa, I'll make you the most advanced amputee the world has ever known."

Again with the promises, I thought. Would I ever learn?

I don't know how such a statement bypassed all the filters in my cautious mind. I have no idea how it leaped out of my mouth in that moment and became a commitment. You'd think I'd have been very cautious after the failure of my first promise. But this new promise seemed to have a mind of its own. It had gotten through my inner filter, and now it was out there. It was my story, and I was going to stick to it.

Neil and Melissa smiled and offered a thumbs-up. They looked at each other and said, "Wow! We've got the right surgeon." Hope is a wonderful *and* a dangerous thing.

Optimism is intoxicating. We sat for a few minutes and talked about prostheses, based on my experience with them. I described the kind I felt Melissa could use. Checking my watch, I realized I needed to keep moving. But as I walked out the door, Melissa said my name, and I turned. "Yes?"

"Dr. Seth, I just want you to know I appreciate everything you've done." Neil nodded in agreement.

I was touched. "Thank you right back, Melissa, Neil. I appreciate the way you've put your trust in me."

Then I walked out the door and started beating myself up for once again raising expectations. Every doctor knows to keep their patients' hopes grounded in reality. And I was good about doing that—usually. Why did I keep promising this particular patient the moon?

I was no authority on prosthetics; I could hardly spell the word. Specialists handled all that, and a part of me wished I'd stressed that a little more.

"Ajay," I told myself as I drove home that night. "You're going to honor that promise. No matter what it takes, you're going to make up for the first strikeout by hitting it out of the park this time. And you'll begin tonight however you can, whatever you can learn."

I pulled up a chair to my computer and googled "*proscetic*." Don't be shocked—even surgeons google medical procedures every now and then.

The first thing that site did was correct my spelling of *prosthetic*. Then it began referring me to articles and updates on the science of artificial limbs—the first, stumbling steps to carry out my promise to make Melissa the most advanced amputee in the world. I grabbed a piece of paper so I could begin furiously scribbling notes.

What was on the first screen? Was it a twenty-first-century, futuristic robotic limb? No, it was a faded pre–Civil War photo of a soldier with a hook for a hand. I might as well have started with the home video of Disney's Peter Pan fighting Captain Hook. So much for "the most advanced amputee in the world." I wasn't about to show Melissa this crude version of an arm.

For some reason I printed out that picture and stuck it in my bag. If nothing else, it would serve as a reminder of the distance between where I was and what I had promised. I had a long journey ahead of me.

23
BECAUSE OF VEGAS

The very next day an e-mail came across my phone. At the time it seemed pretty mundane—a friend reminding me about an upcoming convention. The American Academy of Hand Surgeons was meeting in Seattle in only six days.

My friend's company was always developing new devices and techniques, and we'd recently struck up a friendship. They were inviting me to help make presentations at the conference.

I enjoy these meetings, particularly the chance to get together with other surgeons to share experiences and cases. The problem was that the convention was in Seattle, 2,500 miles away—and in only six days. Fly across the country for a three-day meeting? I was already scheduled for operations and office visits with patients on those dates. I really didn't want to go through the headache of changing my schedule, so my answer was trending toward no.

How had I become involved with a tech company and its presentations? That story had begun at another convention, the American Academy of Orthopaedic Surgeons in Las Vegas, February 2015.

Vegas also makes for a lengthy trip, but it has its attractions. My wife had absolutely no objections to me taking a business trip to Vegas, with the express stipulation that she go along as a chaperone.

We'd flown in on a sunny Thursday afternoon, and I'd registered for the courses that interested me. In no time I was catching up with a number of old friends from my residency in Columbus, Ohio. Old friends and new ideas are a great combination.

A key feature at this convention is always the exhibit hall. All the orthopaedic companies set up booths to show their new surgical devices. Hundreds of thousands of dollars are spent on making these presentations eye-popping with all the bells and whistles.

One particular company's exhibit caught my eye because I'd used its devices and had a few ideas on how to improve them. As I approached the exhibit, I saw a surgeon discussing his distal radius plate. This was a device used to rejoin bone fragments in the wrist, allowing the bones to heal in the best configuration possible after a break. Wrist surgery has been radically improved because of this innovation.

After he was finished, the other ten surgeons in the audience gradually moved on as I waited my turn to introduce myself. For fifteen minutes he listened with interest as I shared my ideas about how the plate could become even more effective. After a few minutes he turned and called to his engineers, "Come on over here. Dr. Seth has some good ideas."

The engineers and I hit it off and had a great time talking about hand surgery. I loved the fact they were so open to new ideas about their products. One of them handed me a business card and said, "Dr. Seth, when can you come visit us in Florida? We'll show you around, and you can see some of our newer products."

I liked Las Vegas. I also liked Florida but for different reasons: sun, sand, and nightlife! This was going to be an incredible opportunity to be on the cutting edge of our field. As always, I thought of that graduation challenge I'd once received: *Watch for your opportunity to move the science forward.* These were men who were doing exactly that. How could I say no to their invitation?

What I didn't realize was that something else was at work, long before I met Melissa. Connections were being made that I couldn't have known. I've come to understand this is true all the time for all of us. We never know what small building blocks from today will end up as significant towers tomorrow. Las Vegas seemed like just another convention, but it was a turning-point moment.

Once the convention was over, Kim and I flew home, but I remembered my invitation to visit this company. We set a date of July 5, 2015, as soon as my son's baseball season was over.

The day after Independence Day, I was in Florida. It was a great trip; I could feel myself improving as a surgeon. I also spent time with the engineering team, admiring their new products and innovations. The Cuban and Portuguese restaurants were equally impressive. I was having a ball.

When it was time to fly home, my hosts let me know they wanted to work with me again. I was highly flattered by these gifted scientists. They felt I'd be effective at teaching other orthopaedic surgeons how to use their products. It took me about two seconds to accept their invitation. It would have been one second, but I had to tame my euphoria. I needed to seem professional, though I felt as if I'd been selected to test-drive the next-gen Ferraris.

As we flew home, I thought about what seemed like simple good fortune. I'd chosen to wander down a certain aisle of a convention hall in Vegas, I'd struck up a chance conversation, and now I was being asked to teach other surgeons on the front lines of innovation.

Of course, it was a lot more than simple good fortune; I just didn't know it yet.

As the months passed, I became engrossed in my work: the surgeries, patient visits, and Melissa's case. Now I had this e-mail, and I sadly realized I was just too busy. I hoped the chance would come again.

But my friend continued to e-mail; then he picked up the phone and called me on September 3. "Isn't there some way you can make it out to Seattle?" he asked. "We're counting on you to help make presentations on Thursday and Friday during the conference. We need to go ahead and get you teaching and leading. And, of course, we'll have a free day on Saturday. Come hang out with us."

"Listen, I'd love that too. Since July I've dreamed of doing this. But I have a thick schedule of patients and operations, and just flying out to Seattle from here and flying back—I'd lose the travel days as well as the rest. Also, I don't want to leave a special patient I've been caring for and go all the way across the country while she's still in the hospital."

"I understand. Just keep us in mind. We'd love to have you."

There was also the reality of cost. An airline ticket to Seattle at this late date would run in excess of a thousand dollars, a fee I couldn't justify for a three-day trip.

That night, as we enjoyed a family dinner at my favorite restaurant, I picked up my phone. Just out of curiosity, I checked airfares as we were waiting for our food. I typed in the date, the place of departure and destination, and specified a three-day return trip. As I hit Enter, I anticipated seeing airline prices that were as sky-high as their jets.

The travel site said the price was $250.

That couldn't be right. I checked it again. Six days from now, Canton to Seattle, Wednesday to Saturday. $250.

It was a crazy-cheap price. I still didn't plan on attending because of all the other situations that precluded my going to Seattle. But one item had been removed from my list of excuses. I could still cancel if I needed to, so I hit the Buy button and went on with my meal.

Since that time, by the way, I've checked flights with the same advance notice and a three-day return. Every single time I've been quoted in the range of $900 to $1,100. For one day in history, however, it was a quarter of that. Very, very strange.

I finished dinner, climbed into the car, and told Kim, "I just bought a plane ticket to Seattle, and it cost less than the new golf club I bought."

"You bought a new golf club?"

"Never mind."

24

LET'S MAKE A DEAL

Such a beautiful Saturday morning: Labor Day weekend, summer finally fading, temperatures in the high seventies that afternoon, and everyone talking football.

I figured it would be a nice time to make my rounds a little earlier, then try to relax and enjoy the day. I knew that once autumn was in full swing, relaxation wouldn't be on the list of options.

Melissa was heading for Day 20 of her hospital stay, with a fever nearly every day. I hated reminding her that she wasn't going anywhere until it was back to normal. But as usual, she accepted the news graciously and wanted to know what was new in my world.

"Seattle, Washington," I replied. "I was invited to attend a conference there and help with presentations, but I don't think I'm going."

"Why not? What kind of conference?" Maybe Melissa had been in the hospital so long that almost any scrap of information could be interesting. A convention of hand surgeons didn't seem like an engrossing subject. I described the trip and the demonstrations I'd do with my new friends—if I went.

"Oh, you ought to go, Doc! That sounds like something you'd really enjoy."

"Sure. I've never been to Seattle. I like travel, meeting people, learning . . ."

"I think you *need* to relax, and you could do that, too, right?"

"Of course, but duty calls. I'm stacked up with work and appointments. Besides, there's no way I'm flying thousands of miles away from you while you have this fever."

"I'm a big girl. You don't need to babysit me."

"I appreciate that, but let's say something comes up while I'm gone—do you want to have to break in some new doctor you don't even know? After you've got me so well trained?"

She ignored all that and said, "What would it take to get you to go to Seattle?"

"Well, *you'd* have to do it. By getting released from the hospital."

"Seriously?"

"Seriously."

She thought a minute. "What about an airline ticket this late?"

I laughed. "Actually, I've got one. Cost me only two fifty."

"No way!"

"Believe it or not, I got it just yesterday for that price. I have no idea how."

"Okay, that tells me you're *meant* to go. It's a sign. So how do I get out of this hospital so you can go?"

"Well, let's put it this way. Your fever would have to resolve for twenty-four hours—then you'd go home, and I'd go west."

"Again—if I can bring down my fever and get out of here, do you promise you'll go to Seattle? No take-backs?"

"No take-backs. Matter of fact, I'll make a pinky promise."

I offered my little finger, and we sealed a solemn agreement. I had no idea why she was so intent on me making the trip, but it didn't really matter—her fever hadn't budged for her entire stay, and now she was going to *make* it go away? Sure—I figured that was the easiest bet I'd ever made.

"Okay!" Melissa said. "Here's how it will play out. You'll come in on Tuesday morning, early, and see that I have no fever. Wednesday, same thing—come in early, then you'll check me out of here because I won't have a fever either day. Then you have to keep your word and fly to Seattle."

"Just like that, huh?"

"Just like that. By the way, what time's your flight?"

"It's at 4:45 p.m. eastern standard time."

"Perfect! That gives you time to go home and pack a bag."

I shook my head. "You know you're crazy, right?" I had a good laugh over the whole thing.

Labor Day evening came. I made my rounds late so I could visit her last at the hospital. Her condition was unchanged. "Melissa," I said, "your mind-over-matter plan to get your fever down through sheer willpower doesn't seem to be working."

"Just wait." She smiled. "I haven't started yet. Remember, I said it would happen at six o'clock tomorrow morning."

"So your fever has to go down in the next twelve hours. Vegas wouldn't give us good odds on that happening."

I wished her a pleasant evening and headed for the door. Just before I left, Melissa said, "Hey, Doc. Early tomorrow morning, remember?"

"Absolutely." I smiled and rolled my eyes as I headed down the hallway. *You have to admire her spirit*, I thought. *I've tried every solution, yet she thinks she can succeed where medicine has failed. I just hope she doesn't get her hopes too high.*

◆◆◆

On Tuesday morning my alarm went off early—just the way I set it, and just the way I hate it.

Melissa wanted me to be there at six to take her temperature. Deep inside I dreaded reaffirming one more time that her condition was unchanged.

I came to the nurses' station and asked for Melissa's records and vitals. She was riding a twenty-four-day streak with that fever, despite the best efforts of the strongest antibiotics we knew, as well as repeated irrigation and drainage in the operating room. So you could color me skeptical.

I slowly scrolled down the census screen, found LOOMIS, MELISSA, and clicked on the name. My eyes scanned the vitals and focused on the reading for 6:00 a.m. It read:

98.6

I almost shouted out loud and disturbed every patient on the hall. *She'd done it.*

Either that or she'd hired computer hackers to fiddle with our readings. Honestly, I wasn't sure which of these was the least likely. I turned to one of the nurses. "Melissa really, really wants to send me a few thousand miles away."

"I'm sure a lot of us feel that way," laughed the nurse.

I hurried to Melissa's room, and she was grinning broadly as I walked in; she already knew. "Did you see? Did you check my temp?" she asked, though I'm sure my pale demeanor—the look of someone who has just seen a ghost—gave her a clue.

"Equipment must be malfunctioning," I said. "It's trying to claim you're at 98.6."

As I said that, the door opened behind me and another nurse stuck her head in the door. "Did you see it?" she asked. "Melissa's temperature is 98.6."

Melissa laughed.

I shook my head and allowed myself to smile along with both of them. I checked her again, and it came up the same.

"Okay, Melissa," I said. "You win this round, but I'll be back! I'm not packing my bags yet. I still think this whole thing is a little sketchy."

"I'm planning on going home after lunch tomorrow, please and thank you."

"Just don't get too excited. You know what that will do? Send your temperature up again."

I didn't know what exactly was going on, but I was a little worried. Just how disappointed was she going to feel when her temperature rose again an hour, two hours, three hours from now? Still, even this much was crazy. I thought about it off and on during the day as I saw my patients—a nice load of patients on the day after a holiday weekend—and then headed home. There I opened the closet and eyed my suitcase for a moment. *Nah. Not going to happen.* I closed the door.

The next morning I went to the nurses' station to check her numbers. Her temperature and other vitals had been duly measured and logged every four hours during the last twenty-four-hour period.

Six checks. Six readings of 98.6.

I stood there and tried to register what I was seeing. A patient had asked me what it would take for me to change my plans and go to Seattle. I'd said she had to bring her body temperature down to normal for twenty-four hours. She'd done that—*precisely.* Right on the nose and right to the second. Plain coincidence, Melissa's powerful desire, or something beyond either of us had accomplished this feat.

I walked to her room, all but in a daze, and found her preparing to go home—quite calmly, of course. As if she wasn't surprised in the least—just another day. She looked over her shoulder at me and said, "I hate to say I told you so."

I just shook my head. "This time you can say whatever you want."

"No fever for twenty-four hours," she said. "I'm bustin' outta this place after lunch."

Twenty-four days in the hospital with a fever, twenty-four hours with a 98.6 temperature. It was a delight to release her from the hospital, but it was a little frightening too.

She smiled. "And you have a plane to catch, right? Good thing you bought that ticket."

"I guess so. I'm a man of my word, after all. But listen, Melissa." And I pointed my finger at her. "Don't you dare have a fever for the next four days. If I so much as receive a text about your temperature deviating by one-tenth of a degree, I'll walk out of the convention, catch the first flight home, and come drag you back to this room!"

"Oh, that's not going to happen," she said. "I won't be running a fever for the next four days."

"And you'll keep it down the same way you got it down this time?"

"I guess." She shrugged

As I've said, it was a little frightening.

Halfway down the hall, I stopped in my tracks. "I've got to hurry!" I said aloud, and a passing nurse looked at me as if I were crazy. I hadn't packed. I hadn't told my family that I was going. I hadn't sent word to my friends who would be at the conference. Melissa was the *only* one who'd believed I was going on this trip. And, apparently, she believed it pretty effectively.

I drove home, threw some things into a duffel bag, then drove to my office and did my best to explain why I was suddenly on my way to Seattle and why my schedule had to be changed. I sent some text messages, then made it to the airport with only a few minutes to spare. Soon the plane was soaring across the flyover states and toward the blue Pacific.

I put on my noise-canceling headphones and closed my eyes for the long flight. I tried and tried to figure out what had just happened, but like everything else concerning the case of Melissa Loomis and the raccoon bite, we were playing a normal game by somebody else's rules.

Nothing to do but enjoy my trip to Seattle. It was great, and it was also a little scary. In science we like our *why* questions to come with clear answers.

25
CHANGE OF PLANS

As I pulled my luggage through the Seattle airport, I realized what a wild day it had been. It had begun at five in the morning in Ohio before taking a wildly unlikely twenty-five-hundred-mile detour. I hadn't expected to be here, and now I was. I needed my head in the game.

I hopped into a cab and rode to the hotel, which was across the street from the Seattle Convention Center. I wanted a good night's sleep before the first lecture series in the morning. My body and Seattle had a three-hour disagreement about the time of day.

I leave for meetings way ahead of time because I get lost in conference centers—it happens every time, and this time might have been my getting-lost masterpiece: three different failed journeys to the correct room. I would have made a terrible wilderness scout.

Eventually I figured out I was on the wrong floor. I made my way to the second floor, where my friends gave me a warm greeting. It's nice to meet those you know in an unfamiliar city, and I began to feel more comfortable.

The first lecture was on distal radius fracture fixation—the method of fixing a broken wrist with a plate and screws. More than two hundred orthopaedic surgeons from across the country were in the audience. I helped the group with cadaveric demonstrations.

Cadavers (corpses) are often made available for educational purposes. Apart from live surgery, cadavers provide the best way to demonstrate proper procedures. Medical students are often a little queasy with this teaching method in the beginning, but in time they get over it.

It was a very detailed lecture. The presentation was finished by midafternoon; then we had a break before meeting up for dinner as a group.

The next day was largely the same, but I decided to explore downtown Seattle on the following day rather than spend extra time in meetings. I'd never been to the "Emerald City," and I wanted to get a taste of its urban life.

As I left the center, I made a call to check with my schedulers back in Ohio. I went through my list of patients with them, answered questions, and made recommendations, and noticed that Melissa's name never came up.

"What's going on with Melissa Loomis?" I asked. "She's still at home, right?"

"Yes. We haven't heard a word from her. But you could check with Tyler."

Melissa was supposed to call Tyler, my right-hand man, if anything changed. I sent him a text and asked, "Any problems? Any fever?"

A text came back within a few minutes:

> Stop worrying about Melissa. Enjoy yourself and relax.
> Her fever hasn't gone above 99 degrees. It's all good!

I smiled, appreciating his encouragement for me to relax—a tall order with this case, no matter how many thousands of miles away I was. But another reason to smile was that Melissa was at home resting with no fever. Didn't this have to mean her infection was gone—that we'd somehow slain the invisible monster, or at least that it had slunk back to whatever strange cave it had come from?

And, finally, I could smile because Melissa had gotten out of that cramped, claustrophobic hospital room and into her own home and bed. I imagined Vincent Van Dog and Vivian Sue Dog leaping up and down ecstatically to see her again. That alone had to be good for her morale, her mental well-being, and, therefore, her health.

Yet it was kind of strange: She'd made it clear that her motivation wasn't herself at all—she wanted me to take the Seattle trip. So it needed to be a special one. She also wanted me to relax, but I'm not the type who relaxes for relaxation's sake. I like getting something else done while I'm doing that, such as taking a step forward with my career. That's what working with these people could do.

After dinner I headed back to my room and set my alarm for an eight o'clock wake-up the next morning—pretty late by my standards. I slept like a rock. Then I rose at eight on the dot, without a single tap of the snooze button. My body was still insisting it was *eleven* and wondering what had gotten into me.

But then something strange happened. When I hit the button to turn off the alarm, the phone's home screen didn't come up as it usually did. Instead, I was looking at a syllabus of lectures for the American Academy of Hand Surgeons. This was supposed to be my free day, so I hadn't set the seminar schedule on my agenda. I'd planned on putting in two busy days helping with the presentations, then taking Saturday off. The calendar should have been blank, and the home screen should have come up on my phone.

But I was awake, and I had that syllabus in front of me. I looked it over, and my eyes fell on a topic that made my heart skip a beat.

I read it a second time. I realized that, no, I wouldn't be doing any sightseeing today. The listing said:

8:30 a.m. Room 321: Advancements in
Amputations and Prosthetics

A few months ago my response would have been "So what?" But now it might as well have been in red neon, blinking on and off: *Amputations and Prosthetics*.

I wondered again what this information was doing on my phone screen since I hadn't put it there. All I could figure was that maybe that page was somehow the last one I'd looked at the day before—but if that was true, how had I failed to notice this item? Thirty minutes before that lecture, at the last possible minute before it would be too late, there it was.

My inner loafer spoke up but very weakly: *Tyler specifically said to forget about Melissa and have a good time. Even Melissa herself sent you out here to relax. Dude! You owe it to yourself.*

No way. When it came down to it, I had more interest in advancements in amputations and prosthetics than I did in playing tourist. Not only that, but I was living under a life goal to *move the science forward*. It was a mission: caring for my patients was a mission. Scenic Seattle was no mission, just background scenery.

This meant I had thirty minutes to shower, gather my things, pack for home, check out, and make it across the street for the lecture. And for goodness' sake, *no getting lost!*

It was just past 8:30 as I crossed the street, dodging traffic, and hurried into the building. Somehow my feet carried me directly to the room with no wrong turns, no elevator misfires, no misinterpreted building maps. Even so, I had missed the first portion of the presentation. I scurried to the first open seat and looked up at the PowerPoint screen.

It wasn't as if Ohio State was taking on Michigan in football in this room. Plenty of open seats were all around me, it was the first session of the morning, and there wasn't a lot of energy in the audience.

The first screen I saw read "Targeted Muscle Reinnervation." What did that mean? Maybe this stuff was too theoretical after all.

Wrong. That slide was one more "raccoon" in the story: it didn't seem like much at the time, but it had a sharp bite.

I listened and learned exactly what targeted muscle reinnervation meant, and at that moment I knew things were about to change.

26
TMR

Imagine a town with a nice, busy section filled with residential activity. Then a tornado destroys the buildings, the parks, the whole thing. The roads that led to this community are now dead ends, blocked off to traffic—sad roads to nowhere.

But the town gets funding to build a new community. It may be just the skeleton of what was there before, but still it can provide a new home for the lost community. The road crew reopens the roads. They reroute them to the *new* buildings and sites.

That's what targeted muscle reinnervation (TMR) does for the upper arm after amputation. Instead of roads, think nerves. The old neighborhood was the arm, and the new one? Well, that's where prosthesis comes into the story—a "skeleton" version of an arm. But now we've found a way to get the brain to carry the old signals to new locations through reactivated nerves. Those nerves resurrected. *Rewired.*

That language may sound a little dramatic, but imagine you've lost an arm. Think how glorious would be the idea of using that part of your body again.

For no more than fifteen minutes, Dr. Gregory A. Dumanian of Northwestern University traced for us how such dreams can become reality for our patients. I hung on every word, spellbound.

I smiled briefly as an old picture came to mind: the one of the nineteenth-century man with the hook. That claw-like thing was just a piece of metal attached where the hand would have been. You had to manipulate it using your arm. What a poor replacement for a hand that moved each finger by the simple act of your *thinking* about it.

We take it completely for granted—the brain thinks, the hand does. Through technology, now someone could operate a prosthesis in precise ways with the mind rather than physical manipulation.

It was a matter of rerouting three of the nerves, the most important of the "highways" severed during the amputation. The nerves would now communicate with new muscles picked up by the prosthesis, and the amputee could bend and extend an elbow, open and close a hand, rotate a wrist.

Your brain is an incredible computer, simultaneously operating all the muscular machinery that came with your body. But imagine it operating things that *didn't* come with the original package. What if you could look at five manufactured fingers and flex each one, make a fist, as you thought of doing so? The implications for amputees are astonishing. Hopeful. Even miraculous.

We watched videos. We were shown techniques for surgeons. And the whole time, I imagined what this meant for Melissa, who had shown me the prosthesis on her phone and said, "This is what I want. This prosthesis."

I looked around. Was I the only one in the room who wanted to stand up and shout, as if Ohio State had just thrown an eighty-yard touchdown pass? I looked around, and it was business as usual for many of the others.

As the lecture ended, most of them headed out for coffee, I would imagine. I fought my way upstream, to the front, and got Dr. Dumanian's attention. "Minute of your time?" I asked. "What you just described is extremely important to me."

"Let's walk," he said.

As we walked, I told him Melissa's story—Seth Abridged Version.

"Dr. Dumanian, from what I've told you, would this patient be a good candidate for TMR?"

"No doubt! We're having a lot of success with this procedure. At the very least it could possibly help to ease her phantom pain."

Phantom pain is the result of the dead-end nerves that no longer lead anywhere. Those nerves still connect to the brain via the spinal cord, and the brain still sends messages with tasks for the hand and elbow because these thoughts have become instinctive, reflexive. But the messages are like undeliverable e-mail.

The result is a *neuroma*, a tumor that looks like an onion but is made of scarred nerves. This delivers an unpleasant electrical shock. Melissa hadn't described any phantom pain to me, but I knew she had all the ingredients for it.

If the disconnected nerves were rerouted, rewired through the surgery described by Dr. Dumanian, a pleasant result would be for Melissa's phantom pain to resolve. That alone, as he pointed out, was a good reason to consider this surgery. But there were far more exciting ones.

We began to talk about what I should do next. I was so carried away with enthusiasm that I asked him a rather wild question. "Do you think this is a surgery I could perform?"

"Of course you could. Absolutely."

Maybe this was arrogant of me. I'd known about this procedure for a grand total of twenty minutes! This prestigious Chicago surgeon was on the cutting edge of new technology, and here I was, a small-town doctor, wanting a piece of the action. All TMR surgeries so far had been done at large prestigious hospitals, such as the Mayo Clinic, Johns Hopkins, or Northwestern Memorial Hospital.

For someone like me to be checking into it—well, it was like hearing about rocket science for the first time and asking if you could build a space station yourself.

I didn't know a lot about it. Maybe this surgical procedure required a small army of assistants experienced in microsurgery. Perhaps my hospital lacked the facilities required for rehab. Dr. Dumanian had no idea who I was or what the size of my town might be. He just knew what the surgery required, and he figured I could handle it if I was a typical orthopaedic surgeon.

Yet I'm still amazed he urged me onward when he knew nothing of my credentials. Small-town doctors are a dime a dozen, the way some of the cutting-edge doctors see it.

But I made the big ask. He gave the big answer. It changed my life. Big-time.

As I think about it today, it occurs to me that the courses of our lives, like rivers, are carved out by the questions we ask. And the unasked ones become dead ends—live nerves with no place to go. The biggest question of all is "what if," and I've always been willing to keep that one in front of me.

"Let me write down a name for you," said Dr. Dumanian. "This man is co-inventor of TMR surgery. I suggest you give him a call."

"Thank you. I'll definitely do that. Can you and I talk again?"

"Sure. This November I'll be performing a bilateral TMR on an amputee from Australia." The patient had lost both arms, so I'd have a chance to see the procedure twice. He continued, "Why don't you come along as an observer? Cadavers are helpful, but you need to see how we work as a team with a live patient."

"That would be fantastic. I wouldn't miss it."

"See one, do one, teach one. That's your path to performing TMR."

I remembered that phrase: *See one, do one, teach one.* I loved the way it moved forward from learning to acting to spreading the wealth of knowledge. I came away with his number, the number for the co-inventor, and a very big smile. *And to think I was excited about sightseeing. This changes everything.* I couldn't wait to get home and tell Melissa.

I felt like picking up my phone and thanking it for looking out for me with that seminar schedule. I always wondered why they call them smartphones. Mine had sent me to a life-changing meeting I didn't even know about. That's a *really* smart phone.

Before I reached the elevator, I was calling Tyler and jabbering about the surgery. "Wait, wait! Slow down!" he said. "Can you run that by me at normal speed? TMR what? What's TMR?"

I tried to tamp down my excitement enough to give him the facts. "Wait 'til I get home," I said. "You'll be blown away by the surgery we can do for Melissa."

"Speaking of your patient—guess what?"

"Please tell me she's okay."

"I don't know how or why, but she's *exactly* okay. As in a perfect 98.6. She told me she'd promised to keep it that way while you're gone. She's a unique patient, Dr. Seth."

"Yes, she is. I can't wait to tell her about this surgery that helps her brain send signals in a new way. Seems like her brain is already doing things in a new way! Have you ever heard of a patient thinking her fever down?"

"Not outside of science fiction."

I had no interest in downtown Seattle by this time. I spent the rest of the day looking for lectures on upper-arm surgery. I knew I'd need to grab dinner, get to the airport, and take the red-eye flight back home. I was elated. For the first time I was excited about Melissa's possibilities as an amputee. And the way it all happened was amazing in itself—the cherry on top of the sundae.

I'd been tired when I left for this trip. Now I was filled with new energy. I thought I might possibly fly home under my own power.

27
END OF THE DEAL

"Hey, everybody—I'm home!"

There's no better feeling in the world than opening the front door and seeing your kids come running to hug you and welcome you back. "Was it a fun trip?" they asked.

"You wouldn't *believe* how fun it was. I couldn't even *handle* all the fun."

"I bet Seattle is cool," said Trinity.

"Maybe. Didn't see it, but I saw something a lot cooler. I saw a way to make one of my patients into the Bionic Woman."

Blank stares. "Who's the Bionic Woman?"

"Oops. I forgot you guys missed that century. *The Bionic Woman* was a TV show that was a sequel to *The Six Million Dollar Man*."

More blank stares.

"The Bionic Woman and the Six Million Dollar Man were regular folks who had accidents. Then the scientists *rewired* them. They had surgical implants with all kinds of cool technology. I'll show you."

I got out my phone, tapped over to YouTube, and quickly pulled up a clip of Lindsay Wagner as *The Bionic Woman*. Remember before YouTube when all we could do was explain things with words? My kids don't remember that either. Anyway, they watched the clip.

"You're going to give your patient superpowers?"

"Well, not exactly. But when you can't use your arm or leg anymore, then getting a new one is a *kind* of superpower."

"Okay . . . did you bring us any souvenirs?"

It's really, really difficult to impress kids today. They live in a world of wonders, and nothing is beyond the imagination. But that was okay with me for the moment. I was finishing a very long day and a very long flight. It was 9:00 a.m. (EST), and I was ready to sleep for the rest of the day. Which I would have done if my daughter hadn't had a softball game later on. I catnapped and sat in the stands for her game.

Later in the evening, I relaxed with a nice bowl of ice cream and turned on ESPN SportsCenter. As I took my first bite, my phone, which isn't polite about interrupting, let me know I had a text. It was Tyler, and his message was a short one: 103.

It took an extra instant to assign meaning to those digits. Melissa's fever had gone up again!

Which would have been beyond unbelievable for any patient other than Melissa.

The phone showed a time of 8:15 p.m. I sent Tyler a text to send Melissa to my office at 9:00. I wanted to see for myself if her temperature had really jumped to 103. Maybe the last reading was inaccurate.

I clicked off the TV and thought for a minute. Fevers don't recur. When they leave, they check out for good. This was definitely the strangest medical case I'd ever handled, and it seemed to take a new and different twist every week.

Melissa and Neil met me at the office. "I'd have been fine with you keeping that fever down, you know," I said. She smiled quizzically and shrugged.

The office was a dark and empty place on a Sunday evening. I led them to our usual examination room and removed the dressing on her arm. The stump looked very good. The end of it had been left open as an outlet for any remaining bacteria but was almost closed.

"When did the fever start to rise again?" I asked.

"All through the day, just gradually," Melissa said.

"When was the first time you took your temperature and the thermometer showed it rising?"

"At nine in the morning it was 100. That was the first time."

I looked at her. "Are you sure about that?"

Both of them nodded. "Nine this morning."

"In other words, the moment I arrived home from my trip."

She laughed when she realized what that meant. Neil looked much more concerned.

I said, "I thought we'd gotten past this fever. That *has* to happen for you to get down the path to complete recovery."

"You only said I needed to get my fever down for four days," she said. "Enough for you to leave and then to stay for the whole conference. I did what you said. I guess when that period of time was over, I just let my body allow the fever to return."

They never prepare you for stuff like this in medical school.

"Melissa," I said. "You don't have to take everything I say so literally."

"I try to follow directions precisely." She was smiling, amused. I guess this was better than having her upset about her fever returning. Especially with what I had to tell her now.

"You know what this means," I said.

"I need to go back to the hospital."

Lots of my patients would have cried, shouted at me, or just put their heads down in their laps and moaned. Many would have lamented over the unfairness of the world. And any of it would be understandable; simple human nature. But not *Melissa's* nature.

Or maybe the strange progress of her fever happened because it *had* to happen. In my mind it had become the one condition left to keep me from going to Seattle. Airline ticket price had been one, too, but that one had also mysteriously (and temporarily) disappeared. The price went down, the fever went down, and I went west. Once I came back east, *both those things rose again*. Along with the airline and the unexplained fever, my phone was also in on the conspiracy—it had directed me to a lecture at the last possible moment.

It seems simple to me. Certain odd things had to happen for me to go to Seattle and into the right lecture. If they hadn't, and I hadn't made the trip, Melissa's life and my life wouldn't have continued on this mysterious journey into something that would reach far beyond the two of us and into an unimaginable new world.

28
"YOU CAN DO THIS"

Melissa arrived at the hospital, and we began a new game of try this antibiotic, try that one. After one week I called in Dr. Al Nemr. We had a serious talk about Melissa's condition and decided to stop the antibiotics.

As always, Melissa was completely comfortable with this strategy. "You're the doctor," she said. "I trust you to take care of me."

Sometimes a patient can get an antibiotic-induced fever. Since none of her cultures produced any organisms, this seemed a possibility—and our last, best hope.

"I'll say this much, Melissa. If your fever spikes up to 120—you're *still* out of here, okay? You've been living off our delicious hospital food for too long."

She laughed, but I was serious about getting her back home. I wanted her in that one place where people are most comfortable: under their own roof. It would be medicine for her to sit in the backyard and watch her dogs run and romp.

We stopped all the antibiotics that next day, and yes, you guessed it, her fever went away. I thought about all the meds we'd injected into her body. All told, she'd been in the hospital for thirty-one days, endured ten surgeries, received twelve pints of blood, and taken twelve different IV antibiotics—and, miraculously, when we stopped all the man-made medicine, her fever was gone.

The next day I pulled out the phone number Dr. Dumanian had given to me for the physician at the Rehabilitation Institute of Chicago. I was ready to start working on a tentative TMR plan for Melissa, pending her approval of it, and the doctor was eager to discuss Melissa's options. For about forty-five minutes we dug into the details of her situation and how TMR might impact it.

"How many times has this surgery been performed?" I asked.

"Depends on who you talk to. Around the world, certainly no more than two hundred times."

"Any TMRs coming up on your schedule?"

"Actually, Dr. Dumanian has one here in Chicago on November fourth. It's on a bilateral upper-arm amputee from Australia."

"He did mention that to me. Bilateral, meaning I could see it done twice?"

"Exactly. Both arms, and a perfect learning opportunity."

Dr. Dumanian had laid out the strategy of *see one, do one, teach one. See two* would be twice as good. "I'll be there," I said. "But you're sure this has been done two hundred times?"

"I think so. Why? Do I detect a trace of disappointment in your voice?"

I took a deep breath. "You see, doctor, I've always had a dream, ever since my fellowship and starting out in practice. I want to be a part of pushing the science forward in some way. This sounded like that kind of opportunity. Still does, I guess, because it's relatively new. What matters is the best care for Melissa, but I was hoping I could *blaze* a trail in the process, instead of following a well-worn one."

"I get that, Dr. Seth. I'm the same way. We want to be part of contributing to the future." He paused, then added, "That said, how would you like to do something amazing?"

"Always."

"Get your patient to feel her prosthetic hand."

"Say that again," I said.

He told me he meant exactly what he said—a patient could possibly experience a sensory response, a feeling, through the prosthetic.

"Listen," he said, "I have a former student. I think he's about fifty miles away from you, in Cleveland. He's a researcher at the Cleveland Clinic, and he's part of a program working to do what I just said—help people actually *feel* a prosthetic hand. The body would incorporate the new arm as though it has recovered the old one."

"That's—that's *incredible*."

"You bet it is. Now here's the thing. He's been working hard on this procedure with another physician—*but* it's never been done in the United States."

My heartbeat accelerated.

"And there you are, just a few miles down the highway from him," he added. "He'll turn cartwheels if you can work with an amputee who's open to trying this."

This was becoming too exciting. I knew it was time to lay my cards on the table. "I need to tell you, I'm not working at a large university with research grants. I'm just a small-town surgeon—"

"Doesn't matter," he cut in. "Doesn't matter where you practice. If you're determined to help your patient, you can do this surgery."

Those were the words I needed to hear. I took them, held on tight, and never let go.

Of course, I needed Melissa's okay; this was her arm and her future. This had to be entirely and objectively her decision. I now had enough information to introduce the opportunity to her, and she was doing well at home. So I called her and set up a conference.

Melissa, and only Melissa, had the right to decide what would happen to her body.

29

LET'S CHANGE THE WORLD

The meeting that would decide Melissa's future plans was held in the conference room of our practice, Spectrum Orthopaedics. Neil was there, of course, and we also invited the rest of the family.

Everyone took a seat around the table. I began describing my trip to Seattle and how I came to attend the lecture that hadn't been on my agenda. Then I talked about TMR surgery and how it worked. As I spoke, I noticed Melissa seemed to have a certain twinkle in her eye, as if she knew exactly what I was about to offer.

I talked about ordinary prostheses and how they work, and the difference between them. I introduced the idea of targeted muscle reinnervation.

"Melissa, if I lift this Styrofoam coffee cup with my hand, it's really my brain that runs the operation. I *think* about picking up the cup so I can drink from it. I send a signal from my brain to my hand: move the cup, clasp it with my fingers, and then grasp it gently so as not to crush the cup and spill the coffee. There's a lot going on there."

Everyone nodded. I continued, "A lot of different nerves are doing the brain's work without me even realizing it consciously. They go out from the brain to operate the different touch functions of the hand. An amputee has to contract the bicep and tricep at the same time to close their hand. With the surgery you can simply think, *Close my hand*, and it will close. No retraining of the brain is necessary."

I talked about an ordinary traditional prosthesis that doesn't connect to the brain, operated by arm and shoulder movements. I showed everyone that old black-and-white picture of the man with the hook. "We've figured things out a little better since that old photograph. Through rewiring you, TMR lets your brain keep doing its old job, just operating the prosthetic instead of your hand. Just like always, you think, *Open my hand*, and your hand opens as the nerves send that message. *Bend my elbow*—it happens. Because you're aware this is new, at first you'd *consciously* think these things: *Open my hand. Bend my elbow.* But after a while it would be like a reflex, almost instinctive, as you did it in the past."

I used drawings and videos to give examples of TMR surgery and people who'd had it, explaining that it was a four- to six-hour surgery involving significant risks.

"I want to be sure you've carefully considered these risks," I said.

"What risks?" Neil asked.

"Reinfection, for one," I said. Everyone groaned. "Another surgery, even with great care, opens the possibility of another infection just as bad as before. Another risk is that the rewiring goes wrong, your body rebels, and you could lose the remainder of that arm, from the shoulder down. What I'm saying is that I could make you worse, and I mean *permanently*."

I described several other risks. Melissa listened patiently, nodding along, and finally said, "I want it. I want to function as normally as possible, and this is the way to do it."

"Melissa, you've listened carefully and thought about it, right? Because you might want to take some time to consider. These risks are real."

"I want it." There was no flicker of doubt in her demeanor. She did understand what was at stake. After all, she'd already been through something close to a worst-case scenario. She'd lost her arm from nothing more than a raccoon bite. Didn't matter. Now she was only concerned with her best-case scenario, and to her, TMR was it.

Well, that was that. It was time to pull back the curtain and reveal the rest of what was behind it—the grand prize.

"Okay. Now let me lay out a further option—something that is brand-new: there's also *feel.* If I touch that coffee cup and it's too hot, I'll *feel* that and jerk my hand back. Your brain sends messages to move, but it receives messages back, through a different nerve, to allow you to feel."

I paused for a second, then said, "Melissa, there's a possibility that you could actually *feel* with a prosthetic hand."

There were audible gasps in the room. Neil said, "How is that possible?"

"It's a step beyond TMR," I said. "It's targeted sensory reinnervation or TSR. I would remap the nerves—the ones that used to run through your forearm to your wrist. We're tricking your brain into thinking that hand is still there. Your prosthetic can transmit sensations: right now, just simple ones. You touch the table with your pinky, and you know it was your pinky, even with your eyes closed."

"This is all amazing," Neil said.

"It's still new and experimental. It's never been done in this country. Only one other surgeon has done it, and that was in Canada. I haven't. Matter of fact, I've never done *any* kind of experimental surgery. I'm a small-town doctor, Melissa, not a world-class surgeon. I can't stress that enough."

Melissa said, "So what? There has to be a first time for everything."

I smiled. "Not everyone wants to be the guinea pig for that first-timer, though. Especially on risky medical procedures."

"You're the doctor we know. You're the one who has taken care of me. And you wouldn't do it if you didn't feel you could do it well. I trust you completely and won't let any other surgeon do it."

I was the one who experienced "touch" at that moment. I was powerfully touched by her trust in me—the whole family's trust, for that matter. Melissa Loomis is an utterly giving person who has an almost childlike faith in others simply because she's so trustworthy herself. Expecting the best is all she knows.

Again I was struck with the thought that no one in the world could be so perfect for the opportunity before us. This whole saga could have happened to anybody, but it had happened to this person, this chosen person. She was just the one to give hope to countless amputees across the world.

"It means a great deal to me that you feel this way," I said. "I want you to know that I haven't even gotten the information on doing the 'feel' part of the surgery, the TSR, yet. And I won't pursue it until I know you've talked it over thoroughly as a family and know for certain you want to do this—that you're willing to accept not only the possibility of something better, but also the real possibility of something worse."

"I get that."

"And if you do decide to move forward, everyone will ask you why you're having it done in this local hospital, instead of Johns Hopkins or some internationally reputable facility. And you should ask yourself that. In the end, Melissa, this is about you and your life—not me, not anyone else. You would reap the benefits *or* the setbacks."

They all nodded, and Neil placed a supportive hand around her shoulder. I said, "That's why I won't accept any answer today. I don't need one tomorrow or the next day or anytime until you're ready to put forward a decision you feel strongly about."

Melissa nodded again, smiled, and said, "Thank you. I don't need any time. Let's make it happen."

I sighed. But she held up her one hand and said, "Doc, I lost my arm. Now I hear about an opportunity that would give me back a little bit of what I lost—and also a chance to offer a gift to the world.

Have you thought about our wounded warriors? Returning war veterans who have lost limbs while serving our country?"

"As a matter of fact, I have," I said gently. "It makes your spine tingle, doesn't it?"

"Think about the hope," said Melissa. I could hear the emotion in her voice. "Think about what we could give back to them."

Obviously she meant it, and it would be useless to ask her to keep considering. I took a deep breath and said, "Well, then—let's do it. Let's change the world."

PART 4
THE OPERATION

30
THE FOOTBALL PHONE CALL

The next day I was headed to Hoover High School for football practice. I attended about once a week, checking on any injuries or medical issues.

On this day, however, my mind wasn't on football—not even close. I was still thinking about yesterday's meeting with the Loomis family. While at a traffic light I tapped in the number for Dr. Jacqueline Hebert, the colleague whom the Cleveland Clinic researcher had told me about.

Dr. Hebert, an associate professor in Alberta, Canada, led the interdisciplinary clinical team that was handling TSR surgery. The team's projects were known as "Bionic Limbs for Improved Natural Control," or BLINC for short.

BLINC? I had to admit, it sounded like some kind of futuristic operation. What was I getting myself into? I loved it!

As the high school stadium came into view, Dr. Hebert answered the phone. To hear better, I pulled into a parking space while we became engrossed in a terrific conversation about the progress of TSR surgery.

"I'd love to hear about your surgical experiences, in regard to time and technique," I said.

"Actually, I'm not a surgeon, Dr. Seth. I'm a physical medicine and rehabilitation physician. We have two excellent plastic surgeons who deal with nerves on a daily basis—they're our hands-on people in the operating room."

The words *two excellent plastic surgeons* gave me another little shot of humility. Did I really want to assert myself into the world of the most talented and elite doctors?

I confided my worries about being inexperienced, and Dr. Hebert reassured me that with something this new, everyone was a newcomer.

As we spoke, there was a pounding sound around me, and several voices shouted, "Hey, Doc!"

"Can I take it for a spin, Doc?"

"Just a short drive!"

"I think I'm losing you," said Dr. Hebert. "Sounds like another line is cutting in."

"No, Dr. Hebert, it's a whole football team! Long story. I'm really sorry, but can you hold for just one second?"

I looked out the window and began making furious hand signs—*Shoo! Go away!*—with my attempt at a menacing scowl and gritted teeth. Team members were surrounding my car, rapping on the windows and hood. These kids absolutely loved my sports car—my one special toy—and they were always after me about test-driving it.

They were wearing pads and helmets, so a doctor's scowls didn't throw much of a scare into them. They laughed and trotted on toward the field as I said into the phone, "Okay, the coast is clear. Dr. Hebert, I'm not in a league with your top people."

"Sounds more like you're in a football league."

I laughed. "Well, at any rate, I hope this surgery is something I can reasonably aspire to."

She quickly reassured me. "I'm sure you can do this. Just be prepared, be patient, and realize we're in uncharted waters here. Be sure your patient knows that too. Someday all of this will be as routine as fitting a prosthesis is now. But we're just opening that door. So far our results on the four we've done are mixed."

"We get that—both of us. Melissa, my patient, has a realistic view of what she faces. I've stressed the risks, but she's like me; she sees an opportunity to be a part of something exciting."

"We'll help you all we can. You're going to get calls from the two surgeons I mentioned, and they'll answer your practical questions, as well as walk you through the procedure, step-by-step."

After we ended the conversation, it was all I could do not to leap out of the car and run a few plays with the team—I had plenty of adrenaline. Everything I had done in operating rooms had already been achieved many thousands of times; logged, cataloged, duplicated. We call them "cookbook" surgeries because they're familiar paths that have been followed countless times by other doctors.

Now I was moving out of the cookbook. I was in the kitchen with ingredients that had never been mixed together, fixing a dish no one was certain how to prepare, that no one had ever tasted. The recipe didn't exist, other than one article describing what these two Canadian surgeons had done.

There was also no way not to be a little fearful. Fear isn't always a bad thing; sometimes it's nature's way of looking out for us. Fear is respect for danger—in a surgeon's case, for the well-being of the patient. That was first and foremost. But I also had both a lifelong commitment to being part of something new and a patient with the perfect attitude and personal constitution to join me. She understood what was at stake, but she was more than willing to be a part of this—she wanted to reach for the stars. In that way we were two of a kind. It was a high-risk, high-reward venture before us, but Melissa wanted the reward of recovering much of the use of a hand. And I wanted it for her.

Then there was the new promise. I'd told her I would make her the most advanced amputee in the world, and clearly here was the one way to do that. I was determined to keep my new promise to a special patient. I was determined not to let her down again.

31

THE WORLD IN OUR HANDS

The human hand is essential to virtually every action in life, large or small. Stop reading for a moment, wave your hand before your eyes, bend your fingers, and think about the incredible, precision, five-pronged tool at the end of your arm. And your body came with two of them! There's a reason we call someone's key assistant his "right hand."

It's a perfect design: four fingers, each one slightly different, and a thumb. That thumb is *opposable*—it can be placed against the other fingers, fleshy part to fleshy part, so that your fingers create an adaptable ring. You can grasp a handle, turn a doorknob, throw a ball, create fine art with a brush or chisel, or perform orthopaedic surgery.

We say God has "the whole world in his hands." And he placed his world in our hands.

"That's nice," you say. But if you ever lost one or both hands, you'd quickly realize their importance. That's why prostheses were first developed. As far back as the Middle Ages and surely before, there were crude wooden or iron arms, but they weren't very functional. The hook had its uses, as I learned when I first googled it.

After the First World War, when so many amputees marched home, the split hook was developed and had a bit more practicality. But no manufactured implement could come anywhere close to the versatility of the flesh-and-bone hand. If you lost your arm, you lost an infinite range of actions and abilities.

It was the early 2000s when Dr. Dumanian and his colleague, working at Northwestern University, would nudge the science forward. They asked, "What if we rerouted the remaining nerves? Couldn't we rewire the arm and bring the brain and the prosthesis into a cooperative relationship?"

But nerves are only roadways for messages to travel. Muscles do all the actual work. TMR uses the biceps, triceps, and brachialis muscles to receive these signals. The nerves fire those muscles to function, rather than reaching the dead end where old muscles are missing. In essence, what we're doing is not overly complicated: taking severed nerves, placing them into different muscles, and rewiring them toward new work.

TMR surgery asks the question, "Can these muscles, these nerves, live again?" It offers the future opportunity for wounded warriors and others to find new hope through a medical miracle. TMR has already seen phenomenal success in helping patients operate their prostheses with far better and more precise control. It's the difference between simply picking up a paintbrush and being able to paint a beautiful brushstroke. Imagine my emotions when I found a way to encourage patients toward reclaiming something precious that had seemed lost forever.

I knew I'd be focusing on biceps and triceps, keeping things more basic. If you hold out your arm and bend it at the elbow, your biceps (a group of two muscles) does that work. Straighten it again, and the triceps (a group of three) takes over.

The goal of the surgery is to split the biceps and triceps, using each for different functions. The brain would use part of the biceps to bend the elbow. The median nerve, which closes the hand, would be rerouted to the inside biceps muscle. So if you were to think, *Bend my elbow*, that brain message would be routed to the outside biceps muscle. When you'd think, *Close my hand*, the inside biceps would get that signal.

"Open my hand" is a message carried by the radial nerve, and it's rerouted to one of the triceps muscles. The muscle twitch would be picked up by electronic sensors within the prosthesis, and the hand would begin to open. So an amputee can extend the elbow or open and close a hand just by thinking about it, the way others do.

I've already mentioned phantom pain, the problem caused by dead-end nerves. Dr. Dumanian had suggested to me that one of the possible benefits of this procedure would be the decrease of phantom pain.

I have a theory of my own about the cause of phantom pain. During amputation, we cut off three nerves in the lower arm. I believe these nerves continue to fire (that is, carry their signals) because the brain is still dispatching those messages instinctively. It doesn't know it's sending out undeliverable mail. Nothing is there that can return a "Stop sending us these signals" message. The result is confusion, interpreted by the body as pain.

The result for Melissa was a feeling she described as a shock in her hand (that is, the one that was no longer there), as if it were resting on a 9-volt battery and she couldn't jerk it away—a terrible, ongoing pain to cope with because her brain couldn't stop saying "Extend your elbow" or "Open your hand" as it always had.

To avoid that pain? Now, that would be a true gift.

32

A LICORICE TWIST

My daughter Trinity once asked me to give a talk to her third-grade class at Greentown Intermediate School. I had to think about how to explain the human nervous system in such a way that eight-year-olds could understand it. After all, they were six or seven years away from high school biology. So I used a string of licorice candy.

Kids perk up and listen when candy comes into the discussion. Licorice, of course, is a tight fiber of strands woven together. Human nerves are twisted together in much the same way. And inside the larger nerve are smaller nerves, each with a function—either motor (moving a muscle) or sensory (feeling something). The motor nerves move outward from the brain down those twisty strands, sending directions. The sensory ones are on the way back, bringing feedback: "This is too hot!" or "This cat fur feels nice." These two kinds of nerves flow in and out between brain and muscles twined together.

As I learned about the surgery, I studied the many careful steps to take. I'd begin by opening the large nerve (the fasciculus), with a process that is just like stripping a speaker wire. Then I'd separate all the strands so I could test and identify each one apart from the others.

A special instrument would measure the signal to tell me whether I had a motor or sensory nerve.

The key of the new surgery, the surgery to help Melissa feel, was to identify the different sensory nerves in the bundle of nerves. With licorice, I showed the kids how I would pull out one strand and test it to see what kind of signal it carried. It might connect to the small nerves inside the skin so that when, say, the tip of your finger touched the hot dish, an important message about heat would be generated and relayed rapidly to the brain. Then a motor message could be sent: "Bend your elbow! Withdraw that finger!"

Since there were no more fingers, how could we simulate that to get the feeling process started again? By tricking the brain into thinking the sensory nerve was still on the job. The message now came from the skin inside the arm rather than the fingertips. This was the part of the surgery that had never yet been done in the United States.

There was one x factor, a roadblock to the procedure we wanted to do: the health of Melissa's nerves. They'd been through a lot when that arm was ravaged by infection. Plus, she was diabetic. Diabetics can have nerves that don't work well because sugar can alter the nerves' ability to function.

With two months to go before the surgery, I reviewed all this with Melissa and her husband so many times that they could explain it as well as I could. Newspaper and TV reporters could simply get their answers from the patient instead of the doctor. But I had explained only the TMR part (the more common motor surgery) in detail, rather than the TSR part that was designed to help her feel.

I couldn't explain something when I lacked the information on it. Those two surgeons in Canada still hadn't gotten in touch with me, and this made me very nervous. They were the ones who had worked in the OR. They were the only two people on the face of the earth who could tell me what happened in the details of this surgery.

I'd learned from Dr. Dumanian some of the facts of the post-operative rehabilitation procedure. He told me that most of his patients' phantom pain worsened over the first four to six weeks after the surgery. Then it would begin to resolve, assuming the surgery was a success. Four to six *months* after TMR surgery, the muscle function might return, and the sensory portion might begin to work. So there was quite a suspenseful waiting period. And it would take most of a full year before we could accurately assess how much motor and sensory function Melissa had regained.

I made sure again that Melissa knew the facts. I stressed the hardship of that waiting period, the initial pain involved. This would be an uphill climb with no assurances of reaching the crest. It was going to take courage and a whole lot of patience. But Melissa never wavered in her resolution. She persisted in her desire to move forward, to regain some of what she'd lost, and hopefully to make a contribution to science and for our wounded warriors.

Also, the market offered no prosthesis that would allow her to feel. If we were successful with the TSR, her body would be ahead of the technology. With the right publicity, she would help create the demand for it.

For the foreseeable future, the best benefit she personally could hope for was to straighten and bend her elbow and open and close her hand through thought. And it wasn't just a matter of having to say, "Oh well, we tried, and it didn't work." The risk was real. This surgery could make things *worse.* Phantom pain could be here to stay and even sharper, more uncomfortable.

33

A PAPER PROBLEM

Halloween was an odd day to mark on my calendar. But it happened to be the date I'd be flying to Chicago to observe Dr. Dumanian, at his invitation, perform a TMR procedure.

It was a typical Friday for me, with operations from early morning to 3:30 p.m. At four o'clock I got a call from Dr. Dumanian's secretary. The hospital there needed my immunization records. I couldn't believe it. It was an hour before everyone would go home for the weekend.

When it comes to surgery, a sterile environment is no small issue. Paperwork. Rules. Regulations set in stone. These things could keep me from learning the procedure, and that would keep Melissa from receiving the motor portion of her operation, not to mention the sensory portion. How was I supposed to explain to her that her hopes would have to be put on hold because of some dry, impersonal technicality?

I was exasperated. I had planned this trip for two months, and we had a lot at stake on our end too. I tried everything I could think of in the next few minutes—everyone in the office was on it. But we couldn't get those papers. My hopes of seeing a bilateral surgery were shattered.

His secretary called back. "I just talked to Dr. Dumanian," she said. "Board the plane tomorrow morning."

"I can do that, but I don't know what good it will do. Did he say anything else?"

"No, he just said that in a text to me."

That night I could hardly sleep. I was going to fly to Chicago yet miss the big event. I could read a few articles and talk to the doctors, but it's not the same as being present in surgery and observing with your own two eyes.

I was in the Chicago area on Saturday when I received a text message from Dr. Dumanian:

You're cleared for surgery.
See you Mon. @ 8am. Just don't cough!!

Somehow the doctor, on a Friday night, had come through for us. He'd gotten a clearance, and I could be admitted to observe the surgery. I was so glad I'd taken his advice and gotten on that plane. It was simply another coincidence in the series of unexplainable circumstances—this one important because there's nothing quite like seeing something in real time as opposed to reading about it or even watching a video.

Monday morning I made my way to downtown Chicago. Inside the hospital Dr. Dumanian introduced me to his staff and then the patient—a quadruple-amputee from Australia. You can imagine his excitement about the prospect of being able to move a prosthesis with much less effort.

In the surgical suite the operation proceeded, and I took about two hundred pictures to study carefully later. I had a camera with an unbelievable zoom lens, and I was able to capture even the microsurgery aspect of the nerve connection. I knew that pictures I could reference later would be so much more effective than an attempt to take written notes.

I saw both procedures, left and right, then walked out of the operating room with Dr. Dumanian to talk about my own surgery.

"So now you've seen it, up close and personal," he said. "What are your thoughts?"

"It's like you said—there's a lot going on, but there's no reason I can't do this."

"You know, I'm going to be keeping close tabs on your attempt at the TSR part of things. Helping your patient to feel will be a game changer. I'm very eager to see what kind of results you get. We're doing so well on the motor side of things, but the *feel* part is an open book."

"As you can tell, I'm wide open to any guidance you can give me."

"I'm glad to do it. I know you'll give your patient a thorough warning that she'll more than likely be worse off for four to six weeks after the surgery."

"Believe me, I've been harping on that and also telling her it's a six-month wait before she might begin to see if the nerves did grow into the muscles."

He gave me lots of other advice, and I treasured every bit of it.

That night as I lay in bed, I took a moment to thank God for Dr. Dumanian—for the enormous good fortune of meeting him in Seattle and for the extra time and effort he'd given me. He was one more key link in the chain of events that had brought us to this opportunity to do something thrilling for amputees.

I could only wonder what new surprises lay in store for us.

34
THE CAP QUESTION

It was time to have a detailed discussion with Melissa and her family. I called them together and brought them up to speed on my trip to Chicago. I went over all the areas of concern, including insurance issues and medical risks of the surgery we were considering.

"You've told us about the risks," Melissa said, "and we've said we're okay with that."

"Well, I'm going to tell you again. There's a 95 percent probability you'll experience *more* numbness, tingling, and phantom pain initially. Just from manipulating them, your nerves will be hypersensitive, and this can, and more than likely will, lead to more tingling, numbness, and phantom pain."

I spoke also about my concerns with her diabetes. On both these issues, Melissa and Neil assured me they knew there were risks and concerns, but they wanted to move forward.

I continued, "Then there's insurance. The carrier has to agree to two things: an experimental procedure for you in a local hospital and paying for a prosthetic arm."

"What?" Neil said. "They *have* to pay for the prosthesis. I mean, they can't just refuse, can they?"

"Believe it or not, they can and often do. Amputees don't always receive top-quality prostheses. The cost of one might be more than $150,000."

Eyebrows shot up and a couple of whistles ensued as I gave them that number. "Still . . . ," said Neil.

"I agree with you," I said. "There should be no question about it. We have a number of groups busy with research right now to establish proof that certain prostheses are absolute necessities, not luxuries. It's cruel to deny someone a replacement for a lost limb. We have to be better than that."

Melissa said, "They would just expect us to do without?"

"Not quite. They'll often pay for more basic prostheses—a hook or a simple device. You operate it by manipulating the shoulders. The more advanced type is seen as nonessential. I wonder if insurance executives would say that if *they* lost an arm."

I didn't want to upset the family, so I said no more about my feelings on this issue. Someone with a simple prosthesis has a more difficult time opening a jar, using a knife while eating, or doing any of a million basic functions we all take for granted. These basic functions add up to what we call *life*. As long as there's a way to give an amputee something better, it's our moral responsibility to do so. It's simple decency.

On the same day, I pulled our team together and assigned roles. By this time all my coworkers and assistants felt a connection to the Loomis family, and each was eager to help in any way possible. Our business manager would take on the insurance issues. Even the supervisor of the front desk helped by handling public relations—we expected media inquiries. Everyone had a role.

After the meeting, our business manager pulled me aside for a word. She said, "We need to talk about Melissa's insurance."

"I'm all ears. Nothing's wrong, is it?"

"Not necessarily. Just remember you're wanting a very expensive prosthetic. The insurance company could cover all of it, they could cover some of it, or they could cover none of it and insist on a standard prosthesis."

I said, "Well, if they take that tack, then we won't be doing the surgery because it's *useless* without the special prosthesis. It would be like getting a driver's license, then being told you can never have a car."

"We just have to be ready for any eventuality. But there's a bigger problem: insurance caps."

As most of us know these days, insurance companies set a ceiling, a cap, on how much of a patient's care will be covered. The magic number is usually one million dollars. If your care reaches a million and a half (and that's more within the range of possibility than you may think), then the company has agreed to cover only the first million.

Up to this point Melissa had been through ten surgeries, spent a month in the hospital, a lot of time in ICU, and received IV antibiotics and physical therapy. I realized it was possible she was already right at, or even over, her cap!

"Tomorrow morning, first thing, check into Melissa's insurance benefits," I said. "Let's find out what we're dealing with."

As I drove home through the dark November streets, I sighed and realized I was looking at another sleepless night. Why did things have to be so complicated? This whole episode was just a doctor trying to help a patient who lost an arm. Money and insurance shouldn't stand in the way of something like that. I know I'm an idealist, but I was brought to this country because it was a land of opportunity. I was raised to think that in this nation all things are possible and that people are willing to help one another.

I was beginning to realize the word *experimental* made getting the go-ahead from the insurance people like walking through land mines. If I'm a business executive, and I'm dealing with finances in a world of spiraling costs in every direction, would *I* want to pay for something purely experimental? As I thought about that, I began to feel deeply discouraged.

The next morning a text came in from our business manager:

> Talked to insurance co.
>
> You won't believe this. Call me.

I quickly dialed her and said, "Well?"

"No cap," she said.

"What?"

"Melissa has no cap on her insurance. And get this—they pay *100 percent* for a prosthetic arm."

I was in complete shock. "That's—well, that's wonderful, but how can it be? I've never heard of a policy like that."

"Me neither, and you know I deal with this stuff every day. No medical cap? That's crazy in itself. And when it comes to prostheses—they pay for it all?"

It was as if Melissa had found the one policy in the world designed for an amputee in need of an experimental procedure. As if the policy were *chosen*, like Melissa herself seemed to be.

This was unbelievable—totally unheard of for an insurance company in this situation.

"It's beginning to look like somebody up there likes us," I said, as I tried to take in this news. "The obstacles are jumping out of the way all on their own now."

I told everyone I could about our good fortune, and I got a lot of wide eyes and mouths dropping open—except for one person: Melissa. Her standard response was always, "Oh. Good." Just ordinary stuff. She said, "Okay, so let's get this show on the road."

35

CALLED AT CHURCH

I still had to face the hospital's credentialing board. Along with the legal staff, they had to sign off on any form of experimental surgery. This would be a crucial hearing. I set the date for December 10, 2015. This way we would have Melissa home before Christmas.

I was pretty certain I could make my case convincingly. I told them, "Though this would be classified as experimental surgery, the risks are actually very low. That's because I'm implanting no devices in my patient's body. I'm simply rerouting nerves—and I deal with nerves every day. This is just a new way of working with those nerves."

The board's predictable question was, why wasn't the patient going to a larger hospital?

"One fairly simple reason," I replied. "Melissa Loomis doesn't want anyone else to do this surgery. We've given her that opportunity, but through all these surgeries, she and her family have come to trust me. I know I can do this, and she believes in me."

Later I received word that I had a full green light from the board.

On Thanksgiving Day, which was two weeks before the surgery, the local paper included a four-page spread about Melissa and me. The Cleveland local TV news also picked up the story. They told the full story up to that moment—all the facts other than one angle:

the part about me still having no idea how to do the sensory portion of the surgery. I still hadn't received that phone call from those two plastic surgeons up north.

A week went by, and I was starting to panic *very* quietly. I wasn't making a big deal out of how much I desperately needed to hear from the Canadian doctors. But why wouldn't they follow up? Dr. Hebert told me not to worry when I called her again; she'd remind them.

In the last few days before the surgery, the team and I held a dry run. They'd been prepping as much as I had been. They were professional in every sense. We progressed through each phase of the surgery, even using the actual operating room. We knew exactly how Melissa would be positioned, and how and when the microscope would be brought in.

That microscope was a story in itself: a marvel of technology flown in specially from California. The Zeiss Corporation took care of that for us in a wonderful gesture—you don't want to know how much it cost them to ship it, but the microscope itself costs up to $400,000, so special care is taken when that instrument travels.

Have you ever prepared for something so much and so often that you almost feel *over*prepared? Maybe you studied so long for that math exam that you thought you might shatter your brain and all the numbers would come pouring out. I couldn't think of anything else to do other than ask a lot of questions of the two Canadian surgeons. *I needed to hear from them.*

Then finally it came—at church, of all places. One evening, five days before the big surgery, our church was holding a discussion about some new developments in our youth ministry. My wife and I were in attendance. I remember placing my phone faceup on the floor in case I was needed by the hospital. I set it to vibrate, but I also tried to glance at the screen occasionally. About twenty minutes into the meeting, I looked down and saw a missed call: Alberta, Canada.

"I have to redial this missed call," I whispered to my wife. It was the understatement of the year. I hurried out the back door and walked to the far (and quietest) end of the corridor. I returned the call, and—thankfully—it was answered. "Hi, this is Dr. Seth calling from Canton, Ohio," I said. "I'm the one who is scheduled to perform the targeted sensory reinnervation surgery."

The surgeon was friendly. Without much small talk, he began to describe the surgery, and I was comforted to hear that it was just as I understood it.

TSR is very time-consuming and very tedious, but a careful and experienced surgeon would be able to get through it. "You'll do fine," the surgeon said. "I can tell from your questions that you deal with nerves on a daily basis."

I wanted to tell him I was dealing with quite a case of nerves just waiting for him to call. But I stuck to business. "If you don't mind," I said, "I'd appreciate you walking through each step with me one more time. Since this is my first try, I really want to be thorough."

I asked a few more questions, and we again went through all the steps. By the third time we went through the procedure, I felt the confidence and certainty a surgeon needs. I just had to know this stuff backward and forward.

We said goodbye, I put the phone in my pocket, and I suddenly remembered I was at church. This was a good place for the answer to a prayer, which this phone call was. I'd waited eight weeks for this conversation.

I slid into the seat beside my wife with a big smile on my face, a feeling of relief in my soul. Now I knew how to perform this procedure. There was really no reason to worry. And looking at the religious imagery around me, I figured there'd never been a reason in the first place. Somebody was looking out for me.

I whispered to my wife, "Sorry I missed so much of the meeting."

"What do you mean? You were gone for, like, ten minutes."

"Are you sure about that? It felt like an hour!"

Later I was struck that I hadn't written a single note while going over the procedure by phone. The church was filled with pencils and paper, and I hadn't thought to grab any even though I usually take notes on everything. But it all was deeply embedded in my mind.

The day before surgery, I left a message on Facebook. I changed my profile picture to one of Melissa and asked my Facebook friends to do the same in a show of support for her, as they prayed for a successful surgery and safe passage through it for her. I also asked them to share the message on social media, adding that I hoped to get twenty-five shares.

Ultimately the message was shared more than *two thousand* times, and I like to think Melissa received at least that many prayers.

I tried to let the rest of that day be as routine as possible. Jaideep had an eighth-grade basketball game at the school gym, and I wasn't about to miss it. I never miss his games; I've coached all his teams since he was in third grade. I watched him proudly, the shortest kid on the team by height but the tallest in spirit. It was a good time for me, a chance to get my mind off the hospital world.

Afterward, standing around the bleachers, friends kidded me, asking if I got all my surgical training off YouTube videos. I said, "Actually, this thing tomorrow, I learned it from a phone call."

They laughed, thinking I was kidding.

I laughed even harder.

36
TRUST THE FORCE

I left the gym and went to the hospital for one last check. I met ten members of the team, and we moved equipment into place, adjusted the wonderful microscope that had just arrived, and basically assured one another that we had this. That night brought a quiet dinner and an early bedtime. I knew I could do this surgery; getting sleep the night before was another matter.

As I lay in bed, I thought it was a good time to go through all the steps I needed to perform for the surgery. I know what you are thinking: *Really? The night before?* But I had gone over the steps thoroughly, counted forty-one of them, and now I was able to name them all in order. I was ready to take on the surgery.

For some reason, as I drifted off to sleep, the last thought that popped into my mind was *Trust the Force.* A Star Wars reference, an instruction for Luke Skywalker, who, of course, has an arm amputated in battle at the end of *The Empire Strikes Back.* He ends up with a robotic hand—one that clearly allows him to feel.

I'm not the type who attends that massive pop culture convention known as Comic-Con, or is even a particularly big Star Wars fan, but for some reason that line jumped into my brain. *Trust the Force.* The Force I trusted, of course, was a different one than Luke Skywalker's, perhaps, and I believed God was moving us along all the way. Why should I be nervous? I closed my eyes and slept well.

When my alarm went off at 5:00 a.m., I headed over to our local gym, Powerhouse. A reporter would meet me there, the start of news coverage on this notable day. By 6:45 a.m. I was back at home and dressed for work: tie, dress shirt, nice slacks. My wife took one look at me and said, "Don't change your routine."

I looked at myself and sighed. "You're right."

I didn't dress so formally on an operating room day, and there was no reason to change the formula today, to elevate this to a Big Event in my mind or the minds of those around me. I headed back into the bedroom, changed into jeans and a button-down shirt, and slipped my feet into my leather boots. Then I had a few minutes to sit in my recliner and reflect.

I had a couple of last-minute thoughts. I remembered again the promise I'd made to Melissa. I was going to make her the world's most advanced amputee. But what would happen if I lived out this day and discovered that I could not keep this promise—just like the other promise that hadn't worked out, the one to avoid amputation entirely?

It was up to me only to do my best; that's all anyone can ever do. Besides, I was convinced this wasn't about me or even just about Melissa—higher purposes were at work. As I drove to the hospital, I thought about the significance of Melissa's surgery. A world was waiting for this breakthrough. Soldiers on faraway battlefields and in local veterans' hospitals were desperate to receive hope. This wasn't about me—it was for them.

And I thought again about Jayani, the daughter we lost in infancy: departed but still with us every day. In so many ways she's my inspiration in everything I do. I was absolutely certain she would be with me in a powerful sense on this day. I have no explanation for why I would suddenly think about her so vividly and be so convinced she was with me. Sometimes you just know what you know.

God was with me. My oldest daughter was with me. I was focused and ready to start.

I punched in the code for the door to the pre-op area, waited for it to slowly swing open, and stood for a moment, looking inside. Thirty sets of eyes turned to look at me, a small army. The entire surgical team was in place. Melissa's family was present. Members of the media as well as representatives from different orthopaedic manufacturers whose products would be involved in the procedures were there. Two pastors from my church rounded out the delegation.

"Morning," I said and smiled. "What's with the huge crowd for a simple carpal tunnel?"

There was laughter all around, just for a moment. Then a serious, resolute spirit fell across the room.

I then turned to the most important individual present for this surgery. Melissa lay on a hospital bed, where I gave her a quick hug and an incredulous look. "Raccoon socks? Really?"

"I'm dressed appropriately for the occasion, don't you think?"

"You'll be the best-dressed person in the OR. Are you still feeling good about this?"

"Ready to go. More than ready."

"Well, you know what comes next. I need to draw on your arm." I showed her my pen, then said, "Show me how strong you are—flex your muscles."

As she flexed, I identified the places on her biceps and triceps where I'd need to make incisions and marked them with lines to guide me. But she didn't have much muscle function, even after working with a trainer. I asked her to concentrate, and she closed her eyes and focused. After about a minute each muscle offered me the slightest flicker, and I was able to make my marks.

I stepped away and nodded at our local pastor. A large circle formed around the bed, and the pastor began to pray.

Lord, we thank you for the doctors and the medical science available to us on this day, and we thank you for Melissa and her courage. You're indeed the Great Physician, and we ask your blessings and grace upon this patient during her surgery. Bless the hands of Dr. Seth, and guide them toward excellent work. Bring Melissa safely through this operation, and give her strength in the days to come so that she might serve you in her work and life. We pray all these things in your name. Amen.

I said, "Melissa, it's about time for us to get started. Are you ready?"

She smiled and said one of her favorite lines, "You're going to do all the work. All I have to do is sleep."

"Sort of," I said. "You may be snoring, but your body has some work to do. It has to help me out by avoiding any complications that would put our surgery in jeopardy."

"I'll tell it to behave. Like my fever."

"I'm sure you will." The anesthesia team was now wheeling her toward the pre-op holding area.

She looked over to me one more time, and I said, "Let's go make history."

I headed for OR12, the same room where I performed her first surgery. As I entered the room, I saw at least fifteen people, most of them busy at work. Music is usually played through the speaker during surgery. I had made a special playlist, so I plugged my phone into the stereo, and we were all set.

Jim Guilletto, the nurse anesthetist, told me Melissa was asleep and ready. He deserves special mention for remaining for the entire duration of a lengthy surgery. Melissa's vital signs were paramount, and at a moment's notice we might have to stop the mission. He would regularly confirm for me that the patient was doing fine so I could concentrate on Melissa's arm, rather than worry about her heart rate or some other complication.

We spent the next forty minutes getting the microscope set up properly in relation to her arm. Normally I wouldn't have to attend to this, but I had a very small workspace because Melissa's amputation was so high up the arm. Once I was comfortable with the positioning in relation to the patient, as well as with how to use the microscope, I moved it out of the way.

My next task was to make sure all her bony prominences were protected so there wouldn't be undue pressure on the skin, leading to ulcers. We toss and turn during sleep, but in surgery the body stays in one position. Being diabetic, Melissa was more prone to ulcers. I had to inspect her from head to toe to ensure that all areas were protected. The surgical skin team had done a great job creating a special pad for her.

It was already 9:35 a.m., and we were still preparing Melissa physically. I had the neurological team set wires on her to test her nerves. I used what is called an SSEP machine (somatosensory evoked potential). Its probes would reveal what type of nerve endings I was touching when I opened up the three major nerves. We wanted to distinguish motor from sensory nerves—the inbound and outbound lanes between her brain and her arm. Once I'd identified them, I could move them to their designated places.

Setting up the wires consumed another twenty minutes. After this I had to check Melissa's body again to make sure there were no new pressure points. She had wires all over her now, including the surface of her head, where brain activity would be monitored. Starting there, I inspected every wire. Sure enough, there were new indentations in the skin from taped wires. We simply had to protect her skin. We solved the problem with a small piece of foam between the wires and her skin, but this took a lot more time. It was now 10:15 a.m.

Melissa was set, we were set, and now I needed to connect with the surgical reps, double-checking that they had all the equipment and resources we might need during the surgery. After talking to the family one last time and scrubbing thoroughly, I noticed it was 11:00 a.m. and time for the main event.

37
THE WRONG MACHINE

I got into my gown and gloves and approached the bed where Melissa was lying.

Before I made any incision, I turned and asked the circulating room nurse to write all the steps on a wipe board that was in my view. I wanted to make sure no steps were missed. Unfortunately, I hadn't written down the steps before coming to the surgery, so I couldn't hand her a list. As I repeated them to her, just as I had rehearsed them in my head the night before, she wrote each step on the board. And sure enough, there were forty-one, just as I had counted them as I lay in bed the previous night.

Facing Melissa's bed, I closed my eyes briefly and asked God to bless my hand—just a simple but strong request for his guidance. Then a nurse handed me the scalpel, and I made the first incision on the front of Melissa's arm.

Here we go, I thought. *Eight to ten hours, and for every moment of that I need to be as focused as I am right now.* I needed to be laser focused. I was going to eyeball each step on that board, be certain I hadn't skipped one, and make sure that I performed each step to perfection.

I began by making what we call an adipofacial flap. This was to be placed in between the muscles I split. Making it would be my first step, and my last one would be to place it between the muscles. There were thirty-nine steps between these two.

I constructed the flap and dissected down the arm to the end of the amputation site to find the nerve Melissa used to close her hand—the median nerve. This is the nerve ending that causes so much phantom pain for amputees. As expected, I found a neuroma. There is usually a bulbous end, something like an onion, where the scarred nerves come together and create the pain.

I made a nice, clean cut, removing the neuroma, and saw beyond it a healthy nerve—needless to say, a very good sign. Everything depended on her nerves continuing to function near the amputation line.

It was now noon, and I asked for the microscope. Watching closely through the viewer, I began to dissect the individual nerve fascicles. I did this by peeling back a portion of the nerve and finding the little fascicles inside it. These are the ones I've compared to strands of licorice candy, twisted around each other to create a rope of nerves. I separated each one, counting eighteen strands. They were the size of pencil lead.

Now that I had unraveled the rope, I began to test each strand for function using the SSEP machine. The idea was that as I touched each individual fascicle to the probe, a neurologist would study the waveform it produced. This neurologist, who was in another location, would interpret the reading and tell us by phone whether the nerve was used to move the hand, feel the hand, or both.

As the crucial moment came, and I touched the first fascicle to the probe, I looked at the probe and thought, *This looks different. The one in the picture I showed the team wasn't like this. Well, they're always upgrading all the bells and whistles, so this is probably the latest and the greatest.* I dismissed the observation and stayed focused.

We went through one bundle, testing each fascicle to make sure the probe was working properly. I touched the entire nerve, leaving nothing to chance, then waited for the SSEP tech to talk about it with the off-site neurologist.

Come on, I thought. *Let's have it.* I was eager for the news.

The answer came: "We're not getting any signal there."

I felt my heart skip a beat. Had it all come to this? Had Melissa's nerve function shut down in the upper arm?

"Are you saying her nerve isn't working?" I asked.

"No, just that the probe isn't picking up a signal." An important distinction.

I looked back at the probe, this time a bit closer. "This isn't the silver-ball electrode probe they used in the Canadian surgery."

"You're right. We felt this one would be better."

I sighed deeply. All these weeks of prep, and the tech company changed the plan? I'd been obsessed with leaving nothing to chance, and I'd expected everyone else to do the same.

I said, "Do you remember in October when I chose the probe I wanted and asked you to contact that company?"

"Of course, Doc. We just wanted to use the one we thought would work best."

"Yet here we are, with no reading. I have to know the conduction of her nerves, or I can't tell the motor ones from the sensory ones." I was beyond exasperated. I knew those on my team were doing their best, just as I was. But we were at a standstill. I had the patient anesthetized, the surgery in process, the arm open, and we had a technology issue.

The rep said, "Hold on—I'll be right back."

He was out the door quickly, and to this day I have no idea where he went. We took a break until he returned twenty minutes later with the probe from the picture, the exact one I'd counted on. Moving quickly, he set up the instruments and handed a new sterile probe to the surgical tech. We tried again.

I touched the bundle of nerves to the new probe, waited in the silent room for what felt like an eternity, and heard, "We have a signal."

I let out a long breath—it felt as if I hadn't breathed for half an hour—and felt my confidence surging back. We'd taken a nerve-racking detour, but we were back on the road again.

38

COMPLETION

"Motor . . ."

"Motor . . ."

"Motor . . ."

"Motor . . ."

"Motor/sensory . . ."

As I touched each fascicle to the probe, I heard the identification spoken back to me. Each fascicle had some part in helping Melissa move her hand or experience feeling. But we hadn't found our true target, the one that came back as sensory-only. I wanted to identify that one fascicle that gave feeling to the hand.

Finally, as I touched one of them, excitement leaped from the speaker: "That's it! That's the nerve you're looking for!"

Talk about stress! She had only a single fascicle, out of eighteen in the nerve, that had the single function of giving feeling to her hand. I was ecstatic to find that sensory nerve. However, since there was only one, if I cut that nerve as I dissected it away from the rest of the fascicles up the arm—again, like separating licorice strands—there was no way she would have any chance of feeling her hand. This small nerve was found at the end of the amputated arm.

I slowly dissected it out of the nerve as far as possible up the arm, checking along the way to make sure it was completely alive, through the SSEP. I did this from just above the elbow to the top of the shoulder. At the top of the arm, I carefully laid the fascicle on a small prep pad I'd placed on the inside of her arm. Then I took the remaining portions of the nerve and placed them toward the inside biceps.

The one fascicle that provided feeling to her thumb, index finger, and middle finger was now sutured to a small skin nerve under the skin. If this nerve fascicle grew into the skin, Melissa's brain would interpret that her fingers were being touched even though it would be the inside of her arm.

Now it was time to attend to the ulnar nerve. This one gives feeling to the small and ring fingers. Again I opened the nerve and went through the process with the probe. And again one small nerve fascicle gave her touch. This was also transferred to the inside of her skin.

Things were going so well. I checked the clock and was amazed to find it was 5:30 p.m. This was a good point to take a break, give the family an update, and put a couple of bites of something into my stomach—I hadn't eaten all day. I wasn't particularly interested in eating, but I did need to keep my strength up because we still had much to do.

As I left the OR, the anesthesiologists continued to monitor all Melissa's vitals, making sure she was completely stable. Her body had responded wonderfully well all day, and I gave silent thanks for that small but absolutely vital condition for all we were setting out to do. Melissa's diabetes was also cooperating with us. Perhaps it was one more way heaven was smiling on our team this day.

The family was pleased to hear my report. I encouraged them to continue being patient—we still had a lot in front of us, but we were on track.

At the à la carte line in the cafeteria, I selected chicken tenders and a drink, and then noticed seven people were in front of me in line. I could play the "big-time surgery in process" card and proceed to the front of the line, saving a few minutes. Surgeons love going to the front of the line, though they've never figured out how to do it in rush-hour traffic.

I smiled at the thought and decided not to be "that guy." When things are going so well, the grateful response is humility. I had maybe seven hours of operating time still ahead, but I waited my turn to pay before finishing my meal in about two minutes.

When I was scrubbed and ready again, Melissa had been turned onto her stomach. It was time to work on the back of her arm. My plan was to find the nerve that opened the hand and reattach it to a portion of the triceps muscle.

First, I used the microscope and confirmed that all the muscles were still functioning. We'd placed a small amount of dye in her bloodstream, and it helped us see that the blood was flowing normally to all the muscles in her upper arm. Even after a severe infection, these muscles were going to give Melissa full function.

It was nearly 10:00 p.m. before we completed this section of the procedure and turned Melissa onto her back again. I checked every single nerve and found them all to be intact, right where I'd placed them. Everything was looking good.

One oddity jumped out at me. The median sensory fascicle—the very one we counted on to give her the ability to feel—appeared as if it had been connected to her skin all along, even from birth. We could barely see the sutures I had used to connect it. This was also difficult to explain. Nature itself seemed to be cooperating with us.

We began closing the incisions on the front of her arm, placing sterile dressings. Melissa was still stable with no complications. We couldn't have asked her body to come through in better shape.

I took a deep breath and looked around the OR. We were all mentally and physically exhausted but elated with our progress. Only time would tell if Melissa truly would be able to feel through her hand—it was up to God and biology now. But it had been a long, grueling day, and there was no reason to be anything but optimistic.

It was past midnight when we removed all the wires and checked every bony prominence. Amazingly, Melissa didn't have a single red area, even after all the hours in surgery. We decided to keep her intubated until morning, allowing her to breathe through the tube.

Emerging from the OR, I couldn't have felt better. It was one of those moments in life when you know you must be exhausted, but your blood is pumping, and your spirits are high.

It was time to give her family an update.

39

WHAT YOU DO WITH AN IDEA

The whole family was waiting. Their role of simply holding out patiently for so many hours is an underrated task. Anyone who's ever sat in such a room, thumbing through magazines, staring at game shows on television, will understand. It was the dead of night, thirty minutes into a new morning—with sixteen and a half hours of surgery behind us.

I knew they could see the gleam in my eyes, but I confirmed it in words. "We hit a home run," I said. "Every little thing went as perfectly as it could have gone."

There were shouts of joy and hugs all around.

"Let's all get a little sleep," I said, though I wondered how I could possibly practice what I was preaching. "I'll be back in at nine a.m. to see her. Melissa is still intubated and probably will be until morning. It's a precaution we take with the surgery being so lengthy."

We all went home. As I drove I tried to tell myself one thing: "No matter how good you feel right now, Ajay," I said, "no matter how eager and impatient you might be, it's going to be a good six to eight months before we get our final grade on this surgery."

That's how long it would take for Melissa's nerves to grow out and (hopefully) begin to function once more. Who could wait so long?

The second I woke up the next morning, I called the ICU to check on Melissa. To my surprise, she'd been extubated overnight and was doing well. She was awake, her pain was under control, and she was asking when she could go home.

I ended the call and breathed a sigh of relief. She'd gotten through sixteen and a half hours of surgery, then about eight more hours of recovery without complications. I thought about my day—another surgery at eleven, and nine cases during the afternoon. Why hadn't I scheduled a day off after such a demanding day in the OR? Sometimes I made no sense to myself!

I had a heavy load of cases at the time, and I wanted to get them finished before the new year. It was December, which is always wild in medical facilities because people want their procedures taken care of after they've met their insurance deductibles and before those deductibles reset in January. We do what we can to help them with that.

At 9:00 a.m. I proceeded to the hospital. When I reached the ICU, Melissa's curtain was closed, and I knew a nurse was busy checking her vitals. I had a brainstorm about going to the gift shop to get a bouquet of flowers or balloons for my star patient.

As I browsed the store, the bright white cover of a children's book caught my eye. It said, in huge letters, *What Do You Do with an Idea?* The author of the book is Kobi Yamada.

I certainly wasn't looking for a children's book, and I wasn't even sure why a hospital store would stock a book about creativity. I picked it up and began to read the story of a boy who develops an idea. This idea floats along, following him everywhere. I kept turning pages.

The boy begins to show people his idea. Many of them laugh. Some count out the reasons his idea won't work—the usual stuff. But the idea stays with him until one day it changes. The idea has become reality. The final page of the book leaves us with this thought:

What do you do with an idea?

You change the world!

I hurried to the checkout desk with a big grin on my face about buying this book for Melissa.

She was ready to see me when I got back to the ICU. I placed the book on a chair and leaned over the bed rail. She seemed much the same as ever—not too worried about pain, none the worse for wear after a sixteen-and-a-half-hour surgery. Same old, same old.

After she told me she was fine, I told her all about the surgery, outlining the most important steps. Then I said, "Is your phantom pain worse?" I knew I had to ask because Melissa wouldn't have complained about it on her own. I was worried the surgery might have worsened it.

"It's about 90 percent."

"So 90 percent of what it was?"

"No, 90 percent *gone*."

"You're serious?"

This was yet another surprise from Melissa. I wouldn't have been at all shocked if the pain had *increased* by 90 percent. The surgery might have caused the pain to flare up, given all the manipulation of the nerve endings. I knew that she'd had that eerie feeling of her hand sitting on her chest in a vise grip—knowing quite well there was no hand.

The nerve that relayed sensation was confused and offering old signals that could no longer be accurate. We'd tried various doses of medication, but she'd really been troubled by the phantom pain—and if Melissa admitted that, I knew it must be bad.

"Do you feel your hand now?"

"As a matter of fact, it's open and resting here at my side, and there's very little pain."

I smiled and said, "You never cease to amaze me."

"Really? Why?"

"Because, once again, your body lives by its own rules. This kind of surgery might have done a number on your nerves, Melissa. Has someone ever said you were getting on their nerves? I got on yours for almost seventeen hours yesterday! They should be in chaos today and letting you know about it. If we had a successful surgery for all those hours, *and* you're not feeling much pain—well, we're ahead of the game. We're playing with house money."

She laughed as I grabbed the copy of *What Do You Do with an Idea?* and held it up as if we were in a children's classroom. "Do you like stories?" I asked.

She squinted at me dubiously, then listened as I began to read the story, showing her the pictures as I did at bedtime when my children were young. I knew Melissa had to be wondering what in the world this idea-hatching business had to do with amputation and prosthetics. But I came to the end, closed the back cover, and said, "Melissa—thank you for believing in my idea. Thank you for not laughing at it or telling me all the reasons it wouldn't work."

Then I pulled out a pen and said, "Would you autograph this?"

She smiled and said, "I'll sign the book if you'll let me go home."

We both laughed, and I thought about how remarkable it was for a patient to be ready to move on so quickly after such major surgery.

"One more night," I said, holding up a finger. "All systems are go—or go *home*—but we really want to be sure. I'm not saying we're not ready to get rid of you, but if we can have you for one more night, we'll give you the boot tomorrow. You just have to promise me you're really not in much pain."

"I'm not. I feel good; if I can go home, I'll feel *great*."

When I left the room, I thought about the goals I'd set before the surgery: 25 percent of them could now be checked off as done. Melissa continued to have tremendous vital signs, she felt little pain, and on Saturday morning we rolled her down the corridor in a wheelchair and out the door to freedom. The last thing I told her was, "I'm not saying goodbye because you won't go more than forty-eight hours without me seeing you."

That night, after performing nine surgeries, I crashed on the couch. I closed my eyes and realized again that it would be six to eight months before we knew if we'd really been successful—though I felt certain I wouldn't be disappointed.

I knew in my heart that I was on the way to keeping a bold promise to a patient: to make her the world's most advanced amputee.

However, to be honest, I'd gone back on one statement. I'd bought *What Do You Do with an Idea?* with the intention of giving it to Melissa. She insisted I keep the book because I was just like that little boy: I had an idea that I wouldn't let go of. To this day the book sits proudly on my mantel, reminding me of what you do with an idea.

You change the world.

PART 5
THE BIONIC WOMAN

40
THE WAITING IS THE HARDEST PART

For a case like Melissa's, being released forty-eight hours after surgery was definitely the fast track. Even that wasn't fast enough for Melissa. From the moment she was awake, she was asking, "When can I go home?" Given how many days she'd spent in that hospital through the year, who could blame her?

Before signing off on her release, however, I made her look me in the eye and promise she'd keep her side of the bargain, which came to

- absolutely no driving, and
- absolutely no going to work for two weeks.

She wasn't too happy about this, but I had to lay down the law. It was vitally important that she rest her shoulder. Further, she had to come see me on Monday and confirm that she hadn't done either of those things. If I hadn't insisted, she would have been driving her car to work the moment we set her free.

I made a condition for myself too. Mine was:

- don't get ahead of yourself!

I needed to take things slowly, but it was difficult not to become overconfident. The surgery had gone so well, and—most remarkable—Melissa's phantom pain was better instead of being worse, which was what we had expected. And I still had that sense of emotional buoyancy that came from a simple feeling, deep in my soul, that Melissa and I were involved in something bigger than ourselves—that we were destined to make a breakthrough because God was guiding our way.

That's how I felt, but I needed to be utterly realistic anyway. Things could change overnight, and Melissa's pain could become unbearable.

So I resolved to be patient. We had a long wait before the nerves actually grew into her skin and muscle, and that's when we would really know whether we'd been successful. In short, I needed to be as calm and undemanding as my patient was. One day at a time.

Melissa came in on Monday for a dressing change and a checkup.

"You haven't been driving in any NASCAR races?" I asked.

"Not a one." She smiled, tolerant as always of my kidding.

"And you're not sneaking off in the dead of night to put in a few office hours?"

"I wouldn't think of it."

"Well, as long as you continue to behave, I'll let you stay at home." I inspected the sites of the surgery. "Your incisions are looking great. Would you like to know what we did during the procedure? Because you slept through the whole thing."

"I'd love that. I've heard bits and pieces, but I'd like to hear it all in detail."

We reviewed the steps of the surgery, which I could explain in my sleep by this time. Her eyes grew wide when I told her about the probe to check her nerves and the various extra measures we'd taken to protect her skin from abrasions. Of course, there was no way to put into words the thrill of that moment when we identified the sensory portion of the nerve.

"Now comes a challenge of a different kind," I said.

"There's always some new challenge."

"Which is true for life itself, right? Now it's all about patience—a wait-and-see period for several months. Nerve regeneration follows its own schedule. I can't do a thing in the world to make that happen faster—we just have to go on about our lives and let biology do its thing. The average nerve grows one millimeter per day—one inch per month." I held my thumb and forefinger an inch apart to make the point.

"Something like waiting for hair to grow out."

"Something like that, sure. But your nerves have to grow back from inside to trigger muscle movement."

I told both Melissa and Neil how to care for the incisions, including avoiding showers for a few days. She needed to move her shoulder as little as possible, to avoid traction on the nerves I'd rerouted.

"I work out at a fitness place. When can I get back into my routine?"

"Patience. All of that's out in the future. For now, we have to take care of that region where we did a whole day of work and let it heal."

After the examination I had other patients to see, other cases that needed my attention. There are times when it's easy to become preoccupied with one case, but others need our attention, so we have to move on. I performed my surgeries, examined and consulted with patients, and enjoyed my family life. But all through this time, I was eagerly waiting for that slow growth of Melissa's nerves. And I knew it took every ounce of strength she had to just try to get that flicker of movement.

Melissa came into the office every other day, and she continued to show signs of improvement. She was chomping at the bit to be more physically active.

Two weeks after the surgery I wanted to see if she still had that small twitch in her biceps and triceps: the tiny movement she'd shown beforehand. It had been all the muscle response she could produce—just enough for me to locate the muscles and mark them for the incisions. As I examined her arm, I asked her to see what she could do. I hoped at least there would be something to build on.

"Melissa, show me your strength," I said. "Flex those muscles."

She flexed her biceps, and the stump of her upper arm rose to a ninety-degree angle, as it would for anyone. My breath caught in my chest.

"Is that good?" she asked.

"That's amazing. Let's see what your triceps can do. Drop your arm again." I gently touched the back of her upper arm. She flexed, and the arm returned to her side, in the resting position—perfectly.

I couldn't disguise my amazement. She smiled. "Good enough?"

I said, "Good enough? You kidding? Two weeks ago you couldn't fire your biceps or triceps. Nothing but the tiniest twitch. This is outstanding, Melissa. Did you know you could do this?"

She shrugged. "I had no idea. You told me not to move my shoulder or arm, and I did what you asked. This is the first time I've lifted it."

"Please understand—in two weeks, this just doesn't happen. *Pain* often happens, and you've said you had very little of that. I wasn't just making it up when I said we could have gotten bad results, making things worse. We went into this surgery knowing how much was at stake."

She nodded.

"Yet look where you are. Your nerves and muscles are working together beautifully already—as if there was never a surgery. Your body does astonishing things, and I have no idea how or why."

Later I did some research and talked to other orthopaedic doctors, and none of it shed any light on Melissa's astounding progress. To this day I have no idea, at least from a medical perspective, how she did so well.

Sometimes people explain remarkable medical stories with the power of suggestion or positive thinking. They say it's just the mind healing the body. Maybe, sometimes. But in this case I never told Melissa she'd be healed in two weeks. I never hyped her up for a miraculous cure. That would have been as foolish as it was cruel because the science said it wasn't going to happen. I'd gone out of my way to prepare her mind-set in the *opposite* direction, telling her the best-case scenario was months away.

No, it wasn't the power of suggestion. Greater things, greater purposes were at work here.

I thought deeply about that as I drove home, shaking my head in wonder. Melissa's case offered an undeniable picture of something far greater than mind or science at work. Something wonderful that offered a bright ray of hope for people walking in darkness.

I walked in through the door of my home and told myself, one more time, to put this extraordinary medical story out of my mind. It was time to be a husband and a dad. Every evening we'd have dinner, and I'd ask how everyone's day went. I'd describe my own day, but I tried not to make a big deal out of this experimental surgery and all it meant to me. I just couldn't let my family life be dominated by this thing.

Someday we would talk about the significance of Melissa's case and what made the whole story so amazing. For now, it was important to keep my family life separate.

There were times, of course, when there was media coverage, and I wondered what my family thought about it. I was on a local TV morning talk show, and we set the DVR to record it. The next morning, as I was preparing to leave, I passed Trinity, who was lying on the floor watching television. "Hey there," I said. "Did you check the DVR to watch your dad on TV yet?"

She looked up briefly and said, "Nope. I was busy."

"Too busy for a six-minute segment? To see what your dad looks like on television?"

"Like I said. I was busy! I came home from school, I had to eat a snack, I had to go to dance class, I had to come home and eat dinner, I had to do my homework, and I had to go to bed. I was too busy."

I ruffled her hair and thought to myself, *Well, you're doing it right then, Trinity. This is your time, and you know what's important in your life. That recording will keep. Being a child won't.*

Maybe in the world of medicine my legacy will be this particular surgery. Under our roof, however, I'm just Dad.

41
THE MIRACLE OF A TWITCH

Four weeks after the surgery I decided to test Melissa's motor response—the TMR portion of the surgery. We knew her muscles were firing well, but how were they going to respond in functioning with a prosthesis? TMR was the first threshold, and then TSR—the ability to feel—would be the ultimate question.

I planned to wait four months before doing this test. But the continued surprises changed my mind. Melissa's body was overachieving, rising to meet every new goal ahead of schedule. The worst that could happen would be to fall back to Plan A and keep waiting.

It was a Thursday afternoon, and my office appointments were behind me. Melissa was in Exam Room 3, where we normally met every other day. I opened the door and smiled as I approached her. "How are things today, Melissa?" I asked.

"Just fine," she said.

I asked a few questions and examined her upper arm, then said, "Today we're going to take the next step—or at least try to take it. Remember, the TMR portion should help your brain fire the new biceps and triceps muscles to open and close a hand if a prosthesis was attached. That means you would be thinking and the new muscles would respond."

She nodded.

"Now, don't expect a lot," I continued. "It's still early. The nerves still have quite a bit of growing to do. There's probably one chance in a hundred we'll notice your muscles twitching as you think about opening and closing your hand. But having said that, those numbers are for the typical patient—which you are *not*."

She laughed. "I'm a freak?"

"You're an *overachiever*. You'd have to be in my position to realize how much of an outlier you've been as a patient. We saw that soon after the surgery, when your nerves began to fire so perfectly. Who knows? Maybe you'll have another surprise for me today."

"Okay—well, how do you want to do this?"

"Why don't you lie here on the bed? Then relax your arm completely. Just like you're going to sleep. No tension, just loosen up."

She reclined on the bed, and I said, "Now, close your eyes and try to think, *Close my right hand*, as you would have done a year ago."

Her brain, of course, would still believe that hand to be there, and it would send the "close hand" signal as it always had. The question was what would happen next. I watched her arm, and nothing moved.

"Did you try?"

"Yes. I visualized my hand and tried to close it."

We repeated the exercise three or four times with no results. "Melissa," I said, wanting to move on as positively as possible, "that's fine. Nothing happened *this* time, but it doesn't mean the surgery didn't work. It just means the nerves aren't there yet—which is why we originally intended to wait a little longer."

As I was talking, I caught a glimpse of her biceps twitching. *Did I imagine that?* No, I knew I had a trained eye on these things, and there was the slightest movement. Of course, it could just be her moving her shoulder in preparation to sit up. I placed my hand on the biceps muscle and said, "Melissa, one more time. Close your hand—right now."

There was no mistake; the muscle twitched.

"Let's see the triceps." I moved my hand there. "Open your hand."

Same thing: the muscle twitched.

I was astounded. There was simply no way she could be responding this way so soon!

"You just went to the head of the class, Melissa. Still overachieving. Your body just obeyed your command to open and close your hand months ahead of schedule." I wanted to run out into the hall and dance. Within three to four weeks of surgery, she had motor response. We could now say that the TMR surgery had worked.

"This is wonderful," she said.

"You're right about that! At this point I'm not sure there's anything you can't do, other than maybe grow the whole arm back."

"I try to do everything you ask, but I'm not sure about that one!"

"No worries. Remember, I'm going to make you the most advanced amputee in the world." I was feeling really confident at this point, and I wanted to speak my promise out loud one more time.

She was as confident and upbeat as I was. It was a joy to look on my phone the next day and read her Facebook post about seeing a muscle response for the first time since her arm had been amputated. All her friends celebrated together online. And every other day she came into the office for a visit—good attitude, good checkups.

At the sixth week postsurgery I decided to spend about thirty minutes checking her muscles and evaluating the strength of her contractions. It was still early, given our expected course of several months. But I wanted to see if she had any feeling—that is, if we were really on an accelerated course. Things had gone so well, I thought, *Why not look for results in six weeks that normally would be expected in six months?*

It all happened in front of a video camera. A local newspaper was updating an earlier story, and it wanted to have film available online for its readers. So the result was either going to be dramatic and exciting or "Oh well. Maybe later."

I put a small amount of pressure on the inside of her arm, hoping that Melissa might experience feeling. I'd moved that one sensory nerve there—the one that would help her feel her hand. I took a Q-tip swab and began rubbing it over the inside of her arm.

"Feel anything?" I asked. What she would feel would be her hand, not the actual site on her arm that I was touching.

"No. I don't."

I made circles on the inside of her arm with the cotton swab. "Tell me if you feel anything," I said. But still she didn't respond. I increased pressure over the entire area. "How about now?"

"I'm sorry. I don't."

"That's okay. Remember, we're just doing a very early check." I started to place the swab off to the side, but I hesitated to put it away.

That's when science gave way to the unexplainable, one more time. And a moment followed that I'll never forget, no matter how long I live.

42
WHAT JUST HAPPENED?

Melissa hadn't felt anything. I'd tried several times, and there was no sensation. But there wasn't supposed to be. This was the whole course of medical science, of previous observation and understanding, saying, "I told you so." According to science, we had no reason to be running these tests yet; *of course* she didn't feel anything.

No harm, no foul. Just smile, move on, and try again in a few months when you're *supposed to* hope for some kind of result.

But something wouldn't let me do that. Something said, "Don't discard that Q-tip swab! Try it one more time." Well, why not? We had nothing to lose.

I brought the swab back to the area beneath her arm and gave it a slight rub. Melissa's eyes widened. She had a funny look on her face. "What's the matter?" I asked.

"I felt it."

"You felt what?"

"My thumb. I felt my thumb."

It was one of those moments we describe as *surreal*—a time when we seem to stand apart from the event and look on it as if it's happening to someone else. *Something is happening here,* I thought. *Something nudged me to run this test way ahead of schedule, and something wouldn't let me give up when the results were negative. Once again this isn't just another day and another case.*

I felt a surge of strong emotions, including something a little damp in my eyes. I looked up at Melissa, and the only words I could produce were, "We did it."

We had moved and rerouted a nerve to enable a patient to feel her hand from a different bodily location. This meant a prosthesis could be specially fitted to manipulate that nerve and allow her to feel her hand.

Melissa smiled. I knew she was feeling it, too, in her own way—as calm, steady, and strong as an oak tree.

"Well," I said, "I guess it's time to talk about next steps."

Then I looked back at her for a moment with no idea of what to suggest. We had no information on what our next step was because we were so far ahead of expectations for her progress and because this procedure was so new. In essence we were now ahead of technology—a mind-boggling thing to consider.

"The next step is to be encouraged." I smiled. "And to come in on Monday."

"What happens then?"

"I'll have figured out what the real next step is. Right now, I got nothing!"

She grinned. "I can't wait to tell everybody. I felt my thumb the way it's *supposed* to feel, instead of phantom pain."

"That's what we're after."

We said goodbye, and I gathered my things and walked to the car. *What just happened?* I thought. This was like a math problem that didn't add up. Surgery is performed on X date. We know that nerves grow at a rate of Y. We also know they need to grow Z inches on the inside of a patient's arm. The answer should be set, and it is more than six weeks. So what just happened?

We'd eliminated phantom pain, seen her arm moving freely, helped her brain use new muscles to move a prosthetic hand, and now we were on our way to proving she could *feel* that prosthetic hand. This was way, way more than we could have reasonably asked.

At this point people ask me, "What did you do in the surgery to get that kind of result? What special steps did you take?"

The answer is *none at all.* I was a chef who cooked by the basic recipe, though it's a very new recipe.

Nothing I did in Melissa's surgery allowed for such rapid results. I earned no Surgeon of the Year awards—I'd been told I could perform this procedure as an ordinary, small-town doctor, and that's what I did. By breaking down a long and complex operation to forty-one basic steps, I knew our team could handle the surgery.

My point is that none of this had anything to do with me. It's like standing at one end of a basketball court and shooting for the opposite hoop, more than ninety feet away, when you've never tried that shot before. You can read books on shooting a basketball. You can take your stance and send the ball into the air with the best form, bouncing from your toes to get leg strength behind it, pushing the ball out with just the right technique, right hand dominant, a nice arc on the path of the ball. You can do everything you're supposed to. But when it swishes through at the other end, nothing but net, on that first try, you still say, "I'm not good enough to do that. That had *nothing* to do with my ability."

We'd swished this one, but it wasn't about my surgical skills or the excellent team I'd worked with. The entire saga, from the day that raccoon had put two tiny holes in a woman's wrist, had the imprint of higher purposes. Unlikely things happen every day in this world, but when so many unlikely events happen *in one story*, as they had with Melissa's case, I choose to believe that something more than random chance is at work.

At home I walked through the door with my normal strategy in mind for being at home—keep office stuff at the office; don't let

Melissa's story overwhelm family life. Just be cool. Be Dad, with maybe a cheerful "Hi, honey. Hi, kids. I'm home."

"WE DID IT!" I shouted the moment the door was open. So much for playing it cool. Sometimes the dam breaks, even when you try to play it cool.

Everybody looked at me quizzically. "Melissa will feel a prosthetic hand," I said, pumping my fist in the air. "We're way out ahead of schedule—*way* out! She can feel her thumb after only six weeks!"

I know my son and daughter were thinking, *We have a crazy father.* But what else is new?

Trinity looked at me and said, "Jaideep was bossing me around today." I love my kids for keeping me grounded.

43

A DC CONNECTION

At six weeks Melissa felt her thumb, and we knew we'd broken through. At eight weeks she began to feel her other fingers, as if they were just awakening, telling her they were back. What incredible moments.

As we continued to meet, I'd draw five circles on the inside of her arm. I knew precisely where to place them by this time. If I touched one, I knew she'd feel her index finger; another and she'd feel her ring finger. The nerves were present and accounted for: fire when ready. This was more than I'd hoped for. Remember, this surgery hadn't been done in the United States until now. The goal, in my mind, was that if you tapped the inside of her arm, she'd have the general feeling of her fingers being touched. To individually feel her fingers—you had to call it a blessing.

Melissa was rewired for (someday) a bionic right hand even though we never implanted anything in her body. We had hit our goals and then some.

So what were the next steps? We had general thoughts of getting her involved in whatever research she felt comfortable with. And we had Wounded Warriors in mind; we were certain that her case had tremendous implications for veteran amputees.

Twelve weeks after surgery, during early March, we were still looking around for the right opportunity. The first people we wanted to contact were our armed forces. But how exactly do you go about that? You don't just pick up the phone, dial 1-800-PENTAGON, and set up coffee together at Starbucks. It was an odd situation. We'd done the hard part—a medical miracle—but I'd never stopped to think about how to spread the word to take it to the next level.

Before I could get too far in my thinking, I took a brief detour to Washington, DC, as a chaperone and the trip doctor with my son's eighth-grade class. Those were three long, exhausting days. We left on March 9 at 5:00 a.m. I had climbed onto a bus with fifty kids bouncing off the walls and screaming. I knew then I was in for a wild ride.

On the second evening of our trip, we toured the World War II Museum, intending to eat at a pizza buffet afterward. With all four hundred kids, you can imagine the fiasco—a battle scene in itself. As we waited to load the buses, I was talking to Ester Presutto, a teacher at North Canton Middle School, when I accidentally jostled a man who was standing near me.

"Excuse me," I said, and we shook hands since we didn't know each other. Ester knew him, however, and introduced him as an uncle of one of the students. He lived nearby and had shown up for a family visit. Busy as I was, this was the only time I had conversed with anyone who wasn't from our group. After exchanging pleasantries, I asked, "What do you do here in DC?"

"I work with Wounded Warriors."

He must have seen my eyes become saucers.

Ester, who knew something of my story, looked at both of us and said, "Oh, now *this* is amazing." The man was curious to hear what was up.

"Would you be willing to join us for dinner?" I asked him. "I'd love to share something really exciting with you. It could help give a wounded warrior as close to a functioning arm as possible."

"I was planning on coming along. My pleasure."

Over pizza (and eighth-grade chaos), I gave him the short version of our story. He paid close attention, grasping immediately what Melissa's surgery meant for military amputees. After I'd finished, he said, "You were right. This could be a wonderful thing for Wounded Warriors. Dr. Seth, could you come back to DC sometime so I can introduce you to the right people from our military?"

"You bet! This is exactly what Melissa and I have been looking for, but we had no idea how to make the contact."

"I have another idea. Let me recommend that you go to Capitol Hill and talk to your congressman. That can be a great way to get things rolling, particularly a winning endeavor like this one." As he spoke, he gave me his cell number. We would figure out a time to meet later in the spring.

That trip to the nation's capital had nothing to do with my work as a surgeon; it had been planned a year in advance. And yet it accomplished exactly what Melissa and I needed at that moment—it gave us a connection to someone in "high places" who was connected to Wounded Warriors.

The trip was wild and tiring, three days of wall-to-wall adolescent energy. But I couldn't have been more thankful for the way it set me in just the right place at just the right time.

44

AT WALTER REED

If a tree falls in the forest and no one is there to hear it, does it make a sound? That famous philosophical puzzle gets to the truth of what we were learning: If a big medical event happens and no media is there to cover it, does it make an impact?

We knew we'd made a breakthrough, but we'd made it quietly—a tiny ripple in the pond of public awareness. Local media had covered us, but the story hadn't spread. I guess I'd counted on the "right people" to hear about this surgery and come to us. They'd know exactly what to do next. But it simply hadn't worked that way.

Melissa could feel her amputated hand, but the story wasn't getting traction. And that meant progress wasn't being made to create a larger demand for this surgical procedure. Military amputees needed to know. Surgeons needed to know.

But at least now there was hope. Someone with the Wounded Warriors organization knew, and he'd also pointed us to the strategy of working through representatives in Washington.

As Melissa continued her therapy and made progress every week, I talked to the secretary for Jim Renacci, who represents Ohio's Sixteenth Congressional District. We talked back and forth and set up a meeting during spring break, when I would return to DC with my family. That was the only appointment I was able to set up with anyone on Capitol Hill.

When we got to Washington, I puzzled over how to get around from one place to another. I'd been there in March, and I'd seen how it was one more place for me to get lost. So I rented a bicycle—something I figured was a genius move. I could avoid street traffic and zip around from site to site, making appointments and spreading the word. I knew all about riding a bike.

What I didn't know about was riding a bike *while studying a map on my phone*. Picture a grown man in a suit and tie, leather boots hooked to the bike, holding his cell phone, and pedaling like a maniac.

My appointment with Congressman Renacci was set for 11:00 a.m., and at 10:40 I was pedaling toward his office. I followed the streets well until I realized, when I should have been halfway there, that I was going the wrong way. I had ten minutes before the meeting was scheduled to start, and I was pedaling in the wrong direction. Now I began to panic—showing up late was a bad look for meeting with a congressman.

I stopped the bike, memorized the right path for the rest of the trip, and went into Tour de France mode. I hunched down, bag over my shoulder, and pumped the pedals for all I was worth. I realized that a suit and tie weren't optimal apparel for bike riding, but it was too late to do anything about that. I arrived at the House of Representatives building, racked up the bike, and hurried to Congressman Renacci's office.

I opened the door, panting and sweating, and collapsed onto a sofa. Just then the congressman's door opened and a couple emerged. "Hey, Dr. Seth!" said the man. "What are you doing here?"

Have you ever seen a friend or relative in the wrong context? Right face, wrong place. The man was one of my patients, in the congressman's office—small world! Three weeks ago I'd performed carpal tunnel surgery on him. He showed me his hand and said, "You did a great job! I'm so much better. Hey, can we count this as our follow-up appointment?"

I smiled and said, "I should be surprised to run into a patient right now, but take my word for it, I'm getting to where nothing surprises me. And, yes, bonus points for meeting in DC. You don't have to come back to see me unless you have any new problems."

It felt like a reminder, after all the aggravation of trying to find this office with a bike and almost being late, that somebody up there was saying, "Don't sweat so much. I've got this."

The meeting with Congressman Renacci went well, once I caught my breath. He was excited to hear that something so important had happened in his district and for members of his constituency. He said, "I'm letting my aides know right now to contact Walter Reed Medical Center."

That's the national military hospital, and just like that I had an appointment for the next day. I'd be meeting two surgeons who had performed TMR surgeries but not TSR. Things were beginning to happen quickly.

On the following morning I approached the gate at Walter Reed. A private in military fatigues was waiting to escort me to a parking spot marked with my name. I was taken on a tour. As I walked the halls, it felt surreal to be in a hospital that has helped so many veterans.

Then I met with Dr. Benjamin Potter, an orthopaedic surgeon who specializes in military patients with amputated limbs. He'd done the TMR surgery more than thirty times. I was excited about talking with him surgeon-to-surgeon, hopefully coming away with new ideas and techniques. We introduced ourselves, exchanged pleasantries for a few moments, and then he dropped the formalities. He hurried around the corner of his desk and plunked into the chair beside mine. "How'd you do it?" he asked eagerly.

"How'd I do what?"

"Get her to feel, of course! How'd you get her to feel? We're all wanting to know."

I was taken aback. This was a world-class surgeon; otherwise, he wouldn't have an office at Walter Reed. He was among the elite, and I should be the one eagerly asking him questions.

"You're really asking me how I did it?"

"Of course!"

"Well, let me show you. Do you have a sheet of paper?"

He hurried to get one, and I walked through the steps with him. He was interested in every part of the process and particularly in the results that I described.

"Not only did you try this; you had wild success," he said. "Your patient has feeling, and she's way ahead of all the timetables, right? I want to introduce you to a colleague of mine, Dr. Paul Pasquina."

This second doctor soon came into the room, and he was just as intrigued by everything I had to say about the TSR process and the aftermath. I finally felt I had an audience of people who understood the implications and could make some waves for wounded veterans and other amputees. Both doctors pointed out that since I'd performed the procedure without any implants or surgery involving the brain, we had crossed a boundary into unknown regions of medical science. We were redrawing the map for nerves, muscles, and sensory information.

Dr. Potter said, "There's an international conference coming up in Chicago in May—it deals with amputees and prosthetics. The program is set, but I'm thinking that for something this big, we can squeeze you in for a talk. Are you up for that?"

"Absolutely. Better yet, I might be able to talk Melissa into coming with me."

As I was ushered to the exit, and our visit was coming to a close, I felt deep emotion. I told my guides, "I see all of this, and I know it's all dedicated to protecting the men and women who protect our country. Walter Reed is here because we care deeply about our soldiers. I really appreciate this place and all of you."

As I left, I thought about Dr. Dumanian, who had told me, "See one, do one, teach one." In May I would carry out the last leg of that commission. It was hard to believe I'd done all three, against probability and, quite frankly, my own expectations. It felt good to be part of something so extraordinary, so much bigger than me.

The Chicago conference would be a great opportunity to tell others about this surgery and how it was performed. But it would be an even better opportunity to introduce Melissa—the one who had placed everything on the line.

45
AN ARM OF THE FUTURE

Physicians were flying into Chicago from all over the world for the International Symposium on Innovations in Amputation Surgery and Prosthetic Technology (IASPT). Though I was going to be in the big city with some big names in our field, trust me, I felt like a small minnow surrounded by a couple dozen whales in the deepest part of the ocean.

For Melissa, however, it was just another trip. Whenever I asked her to go somewhere, she'd simply say, "No problem at all." From her perspective, all that mattered was that being there might help amputees, particularly injured soldiers.

The day before we left I marked her arm with five colors of Sharpie pens. I made a bold, colorful point on each location where she could feel a finger. I knew this would be a good visual aid for demonstrating the locations where we'd routed the nerves.

The next day we landed in Chicago and made our way to the conference center. I was excited about making this presentation, and I knew Melissa would enjoy the event as well. But I'll be honest—I was nervous too! This would be the most important presentation I'd ever given.

As we walked into the center, I found a syllabus and began to look at the names of the presenters. They came from Johns Hopkins Hospital, Walter Reed National Military Medical Center, Mayo Clinic, and Harvard Medical School. Beneath those names was my own, a doctor from Spectrum Orthopaedics in North Canton, Ohio. Amazing. Attendees had to be looking at that last one and saying, "What could this guy have to say that's worth hearing?"

Our presentation was coming up within half an hour. I examined Melissa's stub and found that my color points were gone. "What's up with this?" I asked her. "I placed marks here, and they're gone after only a day?"

"I took a shower."

"You took a shower? Are you kidding me?"

"When I'm about to make an appearance in front of a roomful of strangers," she said calmly, "I like to take a shower."

I sighed. "I guess that's reasonable. Let's get you marked up again." We went into a side room, and I got out my markers. By now I could quickly locate the spot for each finger. But this time something odd happened. When I got to the thumb area, I held her arm tightly between my own thumb and index finger so I could reach it. As my thumb pressed down, she made an odd facial expression.

"What's up?" I asked quickly.

"It feels like you're squeezing the end of my finger."

I stopped what I was doing and looked at her. "You mean you're feeling actual pressure?"

"Yeah—like my thumb is between your thumb and index finger."

We were twenty minutes away from our presentation beginning, and we'd made another breakthrough. Before now she'd felt only light touch sensation. Now, for the first time, she was beginning to feel actual pressure—a form of pain.

I laughed and said, "Melissa, why did you wait for this moment to do that? I don't have time to make a slide about it for the audience!"

"I'll try harder in the future to tell you stuff," she said and smiled.

This was another unexpected development. Light touch sensation was enough of a big deal; now we were moving toward *ranges* of feeling. As I spoke to the audience that day, I added on this newest discovery. It added a little immediacy to the narrative—Melissa was gaining ability by leaps and bounds, right up to the minute. We hadn't had time to process what this new development might mean, but it was great for the other doctors to have this new information.

The presentation went well. It's a great feeling to hold the attention of a room packed with fully engaged listeners. I could hear the chairs squeak as doctors and nurses leaned forward to catch the details of the surgical procedure and study the slides. I also showed video portions of the surgery. There was a hush throughout the room.

Afterward we broke out into small groups in different rooms, where the attendees could visit with speakers and talk to the amputees. Melissa and I hosted a room for four thirty-minute sessions. Many of the questions were for Melissa: What kinds of sensations did she experience? What was the phantom pain like, before and after? She was poised and impressive in providing clear and detailed answers.

I was asked, "How did you get it to work?" And, of course, they wanted to know why in the world someone like me, a surgeon in private practice without grant support or a large, experienced staff, would attempt this surgery in the first place.

In each session I answered this question the same way: "It was my goal to honor a promise I made to her. I took this on as a challenge, without fear of failing."

After the small-group period we headed to lunch. One of the amputees caught my eye. He was wearing a black prosthesis while the gentleman beside him held a laptop. I felt a quick surge of excitement: this was *the arm*—the one I'd been waiting to see.

It's known as the Modular Prosthetic Limb (MPL), and it's funded by the US government and the Defense Advanced Research Projects Agency (DARPA). This is the only prosthetic arm in the world that can interface with the nerves to move and feel. So, of course, we needed to try one with Melissa. It was the best way to demonstrate that Melissa truly had feeling. If she could operate the arm and feel through it, we'd have dramatic evidence.

"Excuse me," I said, walking up to the pair. "I'm—"

"I recognize you," said the man with the laptop. "You and this young lady made the presentation about helping amputees to feel. I'm Matt Fifer from Johns Hopkins APL."

I knew that APL stood for Applied Physics Laboratory.

"As you can imagine, I've been looking for you too," he said.

We found a quiet place to talk. Matt had questions about the procedure, and I showed him the places on the inside of Melissa's arm where she could feel her fingers. We enjoyed the conversation, and he said, "I'd love to have the two of you come to Baltimore to visit the APL lab."

Another great opportunity and the next link in the chain: showing how Melissa would function with an advanced prosthesis designed for someone like her. The two of us were eager to make that trip to Johns Hopkins.

We had two more days in Chicago, including several conversations with vendors about prostheses that would work for Melissa (if not as well as the one we wanted for her). She learned about possibilities for getting involved in research for new prosthetic limbs. She also met the physician from Chicago, the co-inventor of TMR surgery, who had some ideas about lighter prostheses that might be functional for small-statured patients. One of the problems with prostheses, particularly advanced ones, is that they can be heavy and burdensome to wear and use.

We were tired but pleased when we got back home to Ohio. I knew that our success was no longer the best-kept secret in medical technology.

46
FITTING LIKE A GLOVE

It was mid-June, time for our visit to the APL at Johns Hopkins. Melissa's father traveled with us by air to Washington, DC. From there, the plan was to drive into Maryland.

For the past two weeks we had talked to the Johns Hopkins researchers to bring them up to date on Melissa's progress. The MPL arm, which we were so eager to try, had been tested only for motor, or motion, up to this point. That wasn't surprising since no one else had Melissa's ability to feel as well as move. During the previous week, the team at the lab had been hurrying to come up with a method for the arm to translate the sensory signal to Melissa's touch points on the inside of her stump.

When we arrived at Johns Hopkins, I couldn't believe how immense the place was. It was like a small city. We had no idea where to go, so I rolled down the car window and asked a student for help. "Where are you trying to go?" he asked.

"The Applied Physics Laboratory."

"Okay. Which building or number?"

"I have no idea."

"Do you have a name of someone?"

"Yes, I do—Matt."

He gave me a blank look, and I realized how unhelpful I was being. It was like pulling into a big city and saying that you didn't know the address, but you were looking for "John."

"See this complex behind me? That's the Advanced Physics Laboratory." The student pointed to a vast layout of offices spanning fifty acres. "We have, like, thousands of 'Matts,'" he said with a grin. "I guess I could introduce you to them one by one."

"Okay," I said. "Let's start with the first building. *Or* I could just call the Matt I want." I held up my phone.

"Good plan," he said, and walked on as I made the call.

The research team met us at the entrance. I was shocked by the security procedures we had to go through. I hadn't thought about it, but this is the center for secret governmental and military projects. But I did have the Bionic Woman with me, so it was only right for life to feel like a Hollywood movie every now and then—not to mention the cameras all around us. Motherboard, a multimedia publishing group interested in future tech developments, was making a documentary on us, and six people from the group were filming and directing.

Melissa and I were accustomed to only a single video camera from our local newscaster, so we had to get used to this kind of attention. Then again, didn't we want to get the word out? We pretended the cameras weren't following us around, and we went about our business. Between the constant filming and the security personnel who escorted us even to the bathroom, we knew we'd entered a strange new world. (In case you're interested, they did wait outside the bathroom door.)

We toured the entire lab, large as it was, and then gathered around a table. Matt was there, along with several other interested staff. For the first hours we told our story for background, and after that I got to the details of TSR surgery. I was beginning to realize I might well spend the rest of my life describing those forty-one steps. And all the while I was thinking, *When will they bring out the MPL?* I wanted to see Melissa try out the ultimate prosthesis.

After I'd done my show-and-tell, they brought out the beautiful piece of technology that is the Modular Prosthetic Limb. I knew that in time it would be adapted and improved, but it was already impressive, with every motion a human arm possesses. Each fingertip had a small pad used for touch.

When Luke Skywalker tried on his bionic hand in the final scene of *The Empire Strikes Back* in 1980, it seemed like pure science fiction. But we're getting there.

I was on the edge of my seat as Melissa took her place on a stool. The lab people attached a small device to her stump, connected electronically to the MPL arm through Bluetooth technology—the same tech that allows you to talk on your phone hands-free while you're driving.

You might have expected the arm to be strapped right on to the stump, but she wasn't ready for that. This MPL wasn't specially made for her, so she would control it remotely. Signals sent through the air would tell us what we needed to know. Besides, this was the only arm that could move and feel, and it was made for a 220-pound man.

Matt and Bobby (another lab official) ran the computer to begin learning Melissa's muscle patterns. They'd say, "Bend your elbow." She would then think about doing that, and the computer would pick up the signal sent through her neural path to her muscles.

"Straighten the elbow. . . . Open your hand. Now close it. . . . Open and close again. . . . Raise your shoulder. . . . Lower it."

Basically, Matt and Bobby were telling the computer, "This signal is the one for that movement. When you get this signal, bend the prosthetic elbow" (or "open the hand" or whatever). Once the computer knew how to read Melissa's mind, we might say, it could send the right directions to the MPL.

Maybe in some other story, in some other saga of scientific research, there would have been setbacks and adjustments. There would have been second and third tries, lots of headaches, and dead ends. But not in this story—everyday life should be as much of a breeze as Melissa and that computer getting in sync. On her first attempt to use the MPL, she could raise the shoulder completely. She could bend and straighten her elbow, and she could open and close her hand.

One hundred percent achievement, first try. There were gasps all around the room as she controlled the prosthesis through each movement. Then came smiles and laughter. From there all we could do was spend the rest of the day testing and experimenting to see just how far Melissa could go with her control of the arm.

The high point of the whole experience came when we asked Melissa to think about moving her fingers. Could she possibly move individual fingers separately, as we can do with our hands? We started with the index finger. "Melissa, bend your right index finger."

One second of suspense—and it moved!

All these things were happening, remember, because of one surgery. We'd had no idea if *any* of it was going to work, much less the possibility of advanced, complex movements such as finger control. To move her hand as a single unit, like a club, would be profoundly better than having no hand at all. But wouldn't the use of individual fingers—presenting a ring finger, pointing across a room, giving a thumbs-up, dipping a pinky into the water—wouldn't these be so much closer to the ultimate goal of truly replacing a hand?

We'd held such goals as aspirations, horizons to focus on for the future, but instead we were getting there with our first steps on this incredible journey.

We kept coming up with increasingly complex tasks: "Pinch your thumb and forefinger together, as if you're saying, 'Okay!'" It was almost like a game, challenging the technology and Melissa's potential as the world's most advanced amputee.

"Now reach out and pinch something using that thumb and forefinger." Done. Her motor abilities were all perfectly preserved, as if her arm had never been through the awful trauma and infection.

Perhaps the only person in the room who wasn't amazed was Melissa, who was extremely difficult to surprise or discourage or astonish. She smiled and enjoyed herself, just as if she operated a robot hand every day of her life.

"The arm does the hard part," she said, then shrugged. "I just think. That's not work."

47

HOW A MIRACLE FEELS

We took a break, then did some media interviews. After that, I began to grow nervous. The motor portion of the demonstration had been a wild success, but now it was time for the sensory test. We were going to see if Melissa could feel through a prosthetic hand. We finally would take this great step. Where would it lead us? We were happy with what we'd accomplished, but if she couldn't feel through the prosthesis, it would be like we'd climbed Mount Everest without reaching the summit.

This moment was our summit.

For the feeling portion, biomedical engineers had created five small vibrating mechanical devices. In only a week they had conceptualized and built these tiny transmitters, an amazing achievement in itself. The devices were the size of hearing aid batteries, each one connected by a wire to a Bluetooth device. Five of these, one for each finger, were placed on the appropriate spots on Melissa's inner upper arm where each of her "fingers" was located.

Let's say the prosthetic thumb touched something. The tiny device would produce a small vibration over the area on the inside of her arm where the thumb's dedicated nerve received feeling. The signal would travel to her brain, letting her know she was touching something. The five tiny devices were held on with adhesive wrap.

Since December 10, 2015, Melissa and I had dreamed about this moment. It was that elusive prize we'd been pursuing for so many months. To move helps to perform a task, which is wonderful, necessary, a part of life; to *feel*, however, is to live, to experience, to know. That's what we wanted for Melissa, for wounded warriors, for anyone unfortunate enough to be deprived of that essential gift, the hand.

It all came down to the next five minutes. My stomach was upside down. Melissa sat on a chair in the middle of the room. The room fell silent as we switched on the computer and the engineers prepared for the test.

In my mind I saw the whole path we'd traveled to get to this moment. Taken one by one, all the steps we made were everyday events that occurred to ordinary people. There was an animal bite, a bacterial infection, an amputation, a series of operations, a conference trip, an unplanned lecture attendance, a school trip to Washington—all fairly simple, innocuous steps, just as the forty-one steps in our surgery were nondescript, taken individually. But joined together, steps go somewhere; they create a journey to somewhere new. And our journey brought us to a place we never could have imagined going in our wildest dreams.

As Melissa's upper arm was once again connected to the tiny devices, I thought back even further—my parents sleeping on a YMCA floor, giving up their comfort across the world so their children's futures would be limited only by their dreams; my years of learning, and my realization so early in life that I wanted to pursue orthopaedic medicine; the challenge I'd been given not to settle for the easy life, the comfortable practice, but to push the science forward.

I thought of my oldest daughter—the one absent in body but so very achingly present just the same—and the day I'd walked into the OR with an absolute certainty her spirit would never leave my side that day.

I thought of the promise we'd made at Disney World to pay our blessings forward by helping others. And I thought of how the Creator of life had honored our desire by showing me a way to help countless others—amputees, soldiers, people who needed fresh hope.

I recalled all the unexplainable events that led us here, the fingerprints of an invisible hand reaching out to touch me through an unexpected schedule on a phone, a man beside a bus, an airline ticket, a mysteriously adaptable body temperature.

I had lived the emotions of a lifetime in these months, and I knew Melissa had too. I felt all these things as we prepared for this culminating moment, when we would see whether Melissa could feel her hand once more, but this time through a prosthesis.

Thinking about it, I realized I wasn't nervous at all. Something bigger than me had brought us to this moment, and that something had no Plan B. Destiny never takes a wrong turn. I smiled and looked over at Melissa, who was just as calm as I was even though it was her life and her future that were being measured here.

On an impulse I turned to Melissa's father, who had come along to support his daughter and had sat silently as scientific terms and techno-jargon filled the air. I placed a hand on his shoulder and said, "Would you like to do the honors this time?"

His eyebrows arched a bit, and he said, "I'm more than happy to help."

The research team showed him what to do. "Just touch the thumb on this prosthesis—right here. When you do that, she should instantly feel it and tell us about it. And then we'll be certain the surgery worked."

Melissa's father nodded and approached the prosthetic arm. Like all the rest of us, he was in a place and performing a task that he had never foreseen in his life. He looked up at his daughter, and I saw their eyes meet. He raised his right hand and touched the thumb on the prosthesis.

A lovely rainbow of a smile appeared on Melissa's face. I always liked her smile, but this one was her masterpiece, and, of course, it infected the whole room because it meant the fulfillment of all our hopes.

"I feel it," she said.

We wanted to celebrate, but the lab tech said, "Now, touch the index finger, sir."

He repeated his simple act with the same success, and then, one by one, for each of the five fingers. Every time, Melissa grinned, nodded, and said, "Felt it!"

The lost hand was back. It was invisible to the human eye, but it could move, and it could feel. It could pinch, it could move each finger separately, and it could let her know when it had come into contact with some other surface. Her right side was alive again. Can you imagine what that means to someone who has lost a hand?

Her father may have been happier than his daughter. He had somehow, through science, touched the hand of his daughter again.

None of us knew exactly what to say; the moment transcended ordinary conversation. But a surgeon from Walter Reed had just the right words: "Melissa Loomis, you're the most advanced amputee in the world."

A promise kept.

God was with us in that room—I knew because I could feel his hand.

EPILOGUE

The end? Hardly. This is only the beginning.

If this were merely *our* story, it would have reached its conclusion. But Melissa and I pursued this surgical adventure with the idea that many other people will benefit: most of them we don't even know, and some are perhaps not yet born.

For me, it wasn't about showing what a small-town doctor could do, though I enjoyed that aspect of it. It was about pushing the science forward as I was challenged to do when I finished my medical education and began an orthopaedic practice.

For Melissa, it wasn't about her own capabilities, as much as she cared about that. It was about the best for other people: for wounded warriors and others who can now have a new hope of leading better lives after an amputation. She endured fourteen surgeries without ever complaining or feeling sorry for herself. An extraordinary vision and selflessness made that possible.

These were the higher purposes that we always had in mind. But there was also a higher *power* involved—God—and that was the twist in the tale that we never saw coming.

From the beginning of the experience, when Melissa's raccoon bite behaved unlike the countless others I'd treated, the path took unlikely turns. She had a bacterial infection that repeated samplings could not identify. She had a fever that, strictly speaking, shouldn't have lingered, then shouldn't have risen and fallen with impeccable timing. That timing sent me on a trip that completely changed the course of this story and, therefore, medical history. Then, as if to remind us this thing was far beyond our explanations, that fever returned.

Once we took Melissa off antibiotics, the fever vanished *entirely*, just as if it had served some mysterious purpose and it was time for us to proceed to the next chapter.

All along the way, we were carried by factors out of our control.

The unexpected phone screen, early on a Saturday morning, that motivated me to wander over to a seminar I hadn't planned on attending—and where I learned how Melissa could become the world's most advanced amputee.

The long-awaited phone call that wouldn't come until I was sitting in a church meeting.

My unlikely, last-minute permission to view a bilateral surgery—an absolute necessity for me—that never should have come through since red tape seemed to have cut off any possibility of it happening.

And just when we weren't sure how to get the word out about Melissa's story, I'd literally bumped into an executive with Wounded Warriors, who opened the way for everything that followed. This had happened on an eighth-grade class trip that had been planned a year in advance.

That is far too many coincidences.

I haven't even mentioned the sixteen-and-a-half-hour surgery that went so flawlessly, and Melissa's unbelievable physical motor and sensory response far ahead of schedule.

Yes, this is a miracle story. Allow me to explain.

If you walk on water, everyone agrees: that's a miracle. It's in plain sight. Yet other miracles aren't so immediate or visual or dramatic because they come in a lengthy sequence—a series of dominoes falling, one by one, until something amazing results.

If one or two dominoes fall, it could be a coincidence. When a long line of them falls, I choose to believe someone set up the dominoes. That's a scientific deduction, by the way.

The more this story played out, the more I felt the gust of a strong wind at our backs, carrying us on toward the hope we longed for. Wind isn't something you can see, but you can feel it, and you can see its results. That's what happened here.

One other thing.

You might feel that these are exciting ideas, and, at the very least, they made for a pretty good story. For the characters in the story, these were exciting events; for the readers, they're just a good story. But I've been thinking about that too. One of the points I've noticed about our story is that it happened to a couple of extraordinarily *ordinary* people. I'm a small-town doctor. Melissa is, in so many ways, the girl next door: a very nice lady who quietly lives her life and who loves her husband and her dogs. If we're talking about God here, it seems we're talking about one who loves doing spectacular things with everyday people.

So maybe the central message of this book touches on *you*. *Your* life. *Your* story. Could it be that there are higher purposes for your life? Could God have you in mind as well? Because, believe me, amputees are a great cause, but they're not the only cause. There are other kinds of people who need a ray of hope. And there are countless ordinary people, just like you, who could do something about that.

How? That's a question we all must answer for ourselves. Just begin, as I did, with the idea that life is about more than comfort. It should be about making life better for as many others as possible—through their health, their faith, their well-being, their living conditions. It should be about finding your own opportunity to step out and make some kind of difference.

I believe God chose Melissa because he saw the heart within her, as expressed in her love for dogs and her service to others.

Maybe you have that kind of heart. If you do, then look out! You might find yourself *rewired*, just like Melissa Loomis. Just like me.

You might find that somebody is taking every element of your life and reassembling it, rebuilding it, rewiring it for a new purpose that is greater than your dreams. It's happening every day.

That's why I say it's just the beginning of the story. Ours—and yours.

ACKNOWLEDGMENTS

This book was both a joy and a challenge to write: a joy because of my hope for a book encouraging you to believe in yourself and accomplish the impossible; a challenge because, to be completely honest, I'm a surgeon and not an author. The following people helped me meet that challenge.

Tom and Janet Clark, former teachers and now friends, who read my first draft two years ago and made thoughtful insights.

Melissa Loomis, for being my best patient and giving us your story. I am thankful for our friendship.

Jayani Asha Seth, our daughter who lives in our hearts. Born on Christmas Day 2004, this gift to our family is also a gift to others as she is my inspiration to help patients like Melissa as well as children with cancer.

John Reister, MD, for reading through the book and giving suggestions to make it better.

Rob Suggs, a phenomenal writer with a God-given talent to help authors. Thank you for adding your creativity and expertise.

Debbie Wickwire, senior acquisitions editor at W Publishing Group, for falling in love with our story.

ABOUT THE AUTHOR

DR. AJAY K. SETH is a board-certified orthopaedic hand and upper-extremity surgeon in private practice in North Canton, Ohio. Dr. Seth earned his undergraduate medical degree from a combined program at the Medical College of Wisconsin. He completed his orthopaedic residency program at Ohio State University and a fellowship in hand and upper-extremity surgery at Allegheny General Hospital.

While at Allegheny General Hospital, Dr. Seth studied under four orthopaedic hand surgeons, learning surgical techniques for the hand, elbow, and shoulder as well as microvascular surgery. He also spent one year on a research fellowship in orthopaedic surgery and biomaterials at Ohio State University.

Most notably, on December 10, 2015, Dr. Seth became the first surgeon in the United States to perform a surgery that would allow an amputee to move and feel her prosthetic hand with her brain. During the sixteen-and-a-half-hour surgery, he rerouted nerves in the patient's upper arm to give her the ability to feel as though the prosthetic hand was her own. Six weeks after the surgery, the patient demonstrated that she could feel and move her prosthetic arm and hand.

After the success of this innovative bionic miracle, Dr. Seth established relationships with Walter Reed Hospital in Maryland and Johns Hopkins University. He is in collaboration with Johns Hopkins University for research and development, demonstrating that Dr. Seth's patient is now the most advanced amputee in the world, as they continue to make advancements to improve the quality of life for amputee patients around the world. With great enthusiasm, he has launched the American Bionics Advancement Project to serve as his research and development department in association with his parent company, Bionic Miracle, LLC.

In his spare time Dr. Seth volunteers for local sports teams and enjoys playing golf, basketball, and baseball and spending time with his family, along with their two dogs.